Introduction to Automotive Solid-State Electronics

by

**AEA Service Division
Automotive Service Industry Association**

Howard W. Sams & Co., Inc.
4300 WEST 62ND ST. INDIANAPOLIS, INDIANA. 46268 USA

Copyright © 1981 by Automotive Service Industry
Association, Chicago, Illinois 60601.

FIRST EDITION
FIRST PRINTING—1981

All rights reserved. No part of this book shall be reproduced, stored in a retrieval system, or transmitted by any means, electronic, mechanical, photocopying, recording, or otherwise, without written permission from the publisher. No patent liability is assumed with respect to the use of the information contained herein. While every precaution has been taken in the preparation of this book, the publisher assumes no responsibility for errors or omissions. Neither is any liability assumed for damages resulting from the use of the information contained herein.

International Standard Book Number: 0-672-21825-9
Library of Congress Card Number: 81-52155

Edited by: *Christina Mikulak*
Illustrated by: *Jill E. Martin*

Printed in the United States of America.

Preface

Most people are reluctant to get involved in something new which looks mysterious and difficult, like solid-state electronics. This happened when four-wheel brakes came along in 1924, automatic transmissions in 1946, alternators replacing generators in the 1950's, to name just a few. Today hundreds-of-thousands of technicians beat flat-rate on them every day. Once you are into these systems, the mystery and difficulty disappears. Electronics sooner or later will be in every vehicle system, just as the nervous system goes through the entire human body.

The purpose of this volume is to de-mystify solid state electronics and make you willing to learn more. Anyone who has done tune-ups and carburetion should have no difficulty understanding solid-state electronics. Please note that this volume is not a manual for repairing or adjusting solid-state electronics. That will come later. This book will show you, for example, how similar in function solid-state ignition is to the mechanical breaker-point system. The same basic function is performed in both cases, only the means of doing it are different.

This book takes you step-by-step from a review of basic electricity through the primary units of solid state electronics—(diodes and transistors) as they apply to the ignition, fuel and emission control systems—to microprocessors and on-board diagnostic units and likely future applications. Once you read and digest this manual you will be ready to go the rest of the way into solid-state electronics—the wave of the automotive future.

Contents

CHAPTER 1

REVIEW OF BASIC ELECTRICITY 7
Electrical Terms—Electrical Circuits—Voltage Source—Ohm's Law—Types of Circuits—Magnetism—Electromagnetism

CHAPTER 2

INTRODUCTION TO SOLID-STATE ELECTRONICS 15
Semiconductor Material—Diodes—Transistors

CHAPTER 3

CHARGING SYSTEMS 21
Battery Construction—Alternators—Voltage Regulator

CHAPTER 4

IGNITION SYSTEM 31
Ignition Circuitry—Spark Plug Operation—Ignition Problems—Ignition Coil—Ignition Demand—Switching Devices—Troubleshooting Charts and Service Data for Electronic Ignition Systems

CHAPTER 5

EMISSION-CONTROL SYSTEMS 45
Combustion in an Automotive Engine—Positive Crankcase Ventilation—Evaporation Control Systems—Air-Injection Pump—Exhaust-Gas Recirculation—Thermostatically Controlled Air Cleaner—Electric-Assist Choke Systems—Idle Stop Solenoid—Orifice Spark Advance Control (OSAC)—Spark-Delay Valve—Catalytic Converters

CHAPTER 6

MICROPROCESSORS, COMPUTERS, AND LOGIC SYSTEMS FOR AUTOMOBILES . 61
Logic and Decision Making—Logic Systems and the Computer—The Microprocessor and the Automobile—Feedback Carburetors—Electronic Spark Timing

CHAPTER 7

ON-BOARD DIAGNOSTIC SYSTEMS 75
Diagnostic Capability—Testing Equipment—System Performance Check—Additional Diagnostic Techniques.

CHAPTER 8

ELECTRONIC SYSTEM DEVELOPMENTS 85
Keyless Entry System—Knock Limiters—Wiper Systems—Antiskid Braking—Trip Computers

INDEX 91

CHAPTER 1

Review of Basic Electricity

The purpose of this chapter is to review the underlying concepts and terms used in discussing electrical and electronic circuits. For some of you, it will simply be a rerun of what you already know. For others, it will serve as a refresher course to remind you of what you may have forgotten, And for the rest, it will serve as a condensed course in the most essential elements needed to understand electrical circuits.

Regardless of your previous experience, be sure you fully understand the principles that are covered, particularly those pertaining to Ohm's law. You will need them for the chapters that follow.

ELECTRICAL TERMS

Half the battle of understanding electrical circuits (including electronic systems) lies in a precise understanding of electrical terms. If you are not sure of the exact meaning of an electrical term, your mastery of electronics will be just that much more difficult. Although we will be encountering new terms as we go on, the three terms introduced next are fundamental for the description of all electrical or electronic circuits. If you understand these terms, the others that will be introduced later should present no problems.

Voltage

The volt (represented by the symbol V) is the unit of measurement of *electrical pressure* or *voltage*. Voltage in an electrical circuit can be compared to water pressure in a pipe. Even with no flow of water, there can still be pressure in the pipe. Remember, pressure is the force necessary to make something flow, whether that something is water or electrical current. Voltage is also referred to as *potential* or *electromotive force*. Thus, a flashlight battery is said to have a voltage or potential of 1.5 V.

Current

The ampere (represented by. A), or amp as it is usually called, is the unit of measurement of electron flow or *current*. An electrical current results when an electrical pressure or voltage is applied to a circuit. When you turn on a faucet, the flow of water is caused by the water pressure within the pipe. Likewise, when a flashlight is turned on, the flow of current (electrons) is caused by the electrical pressure (voltage) within the battery. The symbol for electrical current is *I*.

Resistance

The ohm (represented by the symbol Ω) is the unit of measurement of electrical *resistance* to current flow. Resistance can be considered an electrical "friction" that impedes the flow of electrons. Just as a restriction in a pipe impedes the flow of water, resistance in an electrical circuit limits the amount of current that can flow. In some circuits, such as a starter-motor circuit, we try to keep the resistance as low as possible to obtain the maximum current. In other circuits, such as the primary circuit of an ignition system, we purposely add resistance (e.g., a ballast resistor) to limit the current to some desired value. The symbol for resistance is *R* and the schematic representation is a zig-zag line (Fig. 1–1).

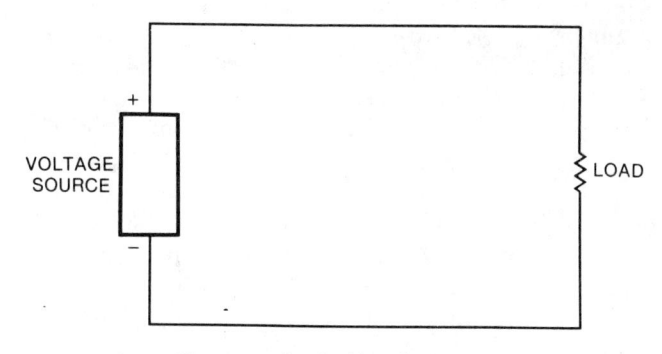

Fig. 1–1. Basic electrical circuit.

ELECTRICAL CIRCUITS

Now let's see how these three terms are interrelated in an electrical circuit (Fig. 1–1). The voltage source could be a battery, an alternator, or some other voltage-producing device. The conductive path

Introduction to Automotive Solid-State Electronics

is normally metallic wire or other material capable of carrying an electric current. In automotive circuits, the vehicle chassis is usually part of the electrical path and is commonly referred to as the "ground" side of the circuit. The load, shown as the zig-zag symbol for resistance, is the device that is to be operated. It could represent a lamp, a motor, a relay coil, or any other electrical device. The resistance symbol is frequently used to represent these various devices (but usually with a label to identify the particular item) because they almost always exhibit some characteristic resistance value.

The circuit must be *complete* for a current to exist. Assume that the circuit starts at the positive terminal of the voltage source. It must continue uninterrupted through the circuit conductors (wires), the load, and back to the negative terminal of the voltage source. The circuit is then said to be complete and current will flow through the load.

VOLTAGE SOURCE

Voltage, or electrical potential, can be created in a number of ways. The most common voltage-producing devices found in automotive circuits are those that create an electric potential by chemical or magnetic action. The battery is an example of a device that produces a voltage chemically, while the alternator is a device that produces a voltage magnetically.

A pump that produces pressure in a water system has an inlet and an outlet. A voltage source is similar in that it has a *positive* and a *negative* terminal. For simple electrical analysis, such as that required in automotive work, it makes little difference which way you consider the current to flow, as long as you are consistent. For our purposes, we will consider current to flow *from* the positive terminal and return *to* the negative terminal. This is called "conventional" current flow. Engineers commonly use "electron" current flow in which the current is assumed to flow from the negative to the positive terminal, this being the true direction of travel of the electrons. You will get the same answers regardless of which method you choose to follow. The most important thing to remember is that the same amount of current flows *from* the voltage source as returns *to* the voltage source.

OHM'S LAW

Now that we have identified the terms and established the basic elements of an electrical circuit, let us consider the relationship of these terms. This is expressed as Ohm's law, the most important and fundamental law in electronics. It is given by the formula

$$E = IR$$

where,
E is the voltage in volts,
I is the current in amperes,
R is the resistance in ohms.

This law states that the voltage *(E)* is equal to the current *(I)* times the resistance *(R)*. It can also be written

$$I = \frac{E}{R} \quad \text{or} \quad R = \frac{E}{I}$$

As you can see, these formulas are simply the original Ohm's law written in a different form. If you know any two of the three factors *(E, I, or R)*, you can always solve for the third.

Ohm's law provides many useful answers about circuit operation and allows us to make an accurate circuit analysis without actually seeing the circuit or making any measurements. Here is how it might be applied in a typical situation.

Assume that we wish to check the condition of the rotor winding in an alternator. The alternator's specifications tell us that the rotor (field) current draw is 2.5 amperes at 12.0 volts. We could connect the rotor to a 12-volt battery, as shown in Fig. 1–2, and measure the current with an ammeter. If it measured approximately 2.5 amperes we could assume that the rotor winding was good.

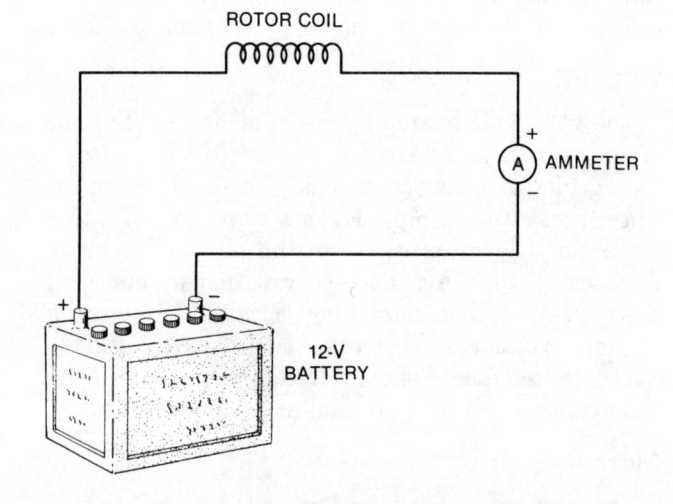

Fig. 1–2. Measuring the current draw of an alternator rotor coil with an ammeter.

Review of Basic Electricity

However, an easier way would be to measure its resistance with an *ohmmeter*. An ohmmeter is a convenient instrument for testing electrical units, particularly when they have been removed from a vehicle. But before we can use an ohmmeter, we must first convert the rotor specification, which was in amperes, to ohms of resistance. In some cases, rotor resistance is specified in ohmic values. But if it is not, we simply use Ohm's law. Since we know the voltage *(E)* and the current *(I)*, we can use the formula $R = E/I$ to find the resistance. Substituting the given values, we have

$$R = 12 \text{ V}/2.5 \text{ A} = 4.8 \text{ }\Omega$$

If the rotor resistance measures approximately 4.8 ohms with an ohmmeter, we can assume that the winding is good. We know, because of Ohm's law, that the rotor will draw the specified amount of current, 2.5 amperes in this case, even though we have not actually measured that current.

The alternate forms of Ohm's law can provide additional circuit information. For example, if a coil has a resistance of 3 ohms, how much current will it draw when connected to a 12-volt battery? In this case, we would use $I = E/R$. Substituting the given values, we obtain

$$I = 12 \text{ V}/3 \text{ }\Omega = 4 \text{ A}$$

TYPES OF CIRCUITS

So far, we have considered only simple circuits; that is, a voltage source and a single load (or resistive element). All other electrical circuits can be classified as either *series* circuits or *parallel* circuits. An example of a series circuit is shown in Fig. 1-3. All of the circuit elements, or loads, are connected in chain fashion. As we mentioned earlier, there is always as much current leaving the voltage source as there is returning to it. This means that the same amount of current passes through *each* element in a series circuit. No matter where a current-measuring meter (ammeter) is inserted into a series circuit, it will always read the same. Now let us apply this bit of information to a typical series circuit and see what we mean by the term "voltage drop."

Voltage Drop in a Series Circuit

The most important fact to remember about series circuits is that *the sum of the voltages across each element equals the source voltage.* This fact is used in testing and troubleshooting automotive electrical systems and, therefore, is a useful one to remember.

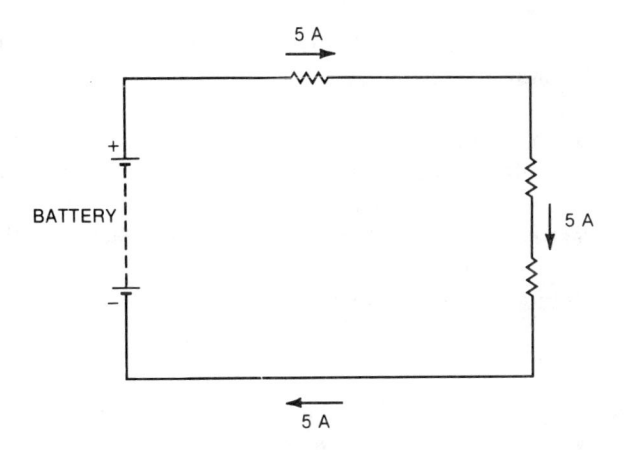

Fig. 1-3. Series circuit. Note that the current remains the same in all parts of the circuit.

Fig. 1-4 shows how this works. Assume that we wish to find the voltage (or voltage drop) across each element in this circuit. The first thing we must do is find the total resistance of the circuit.

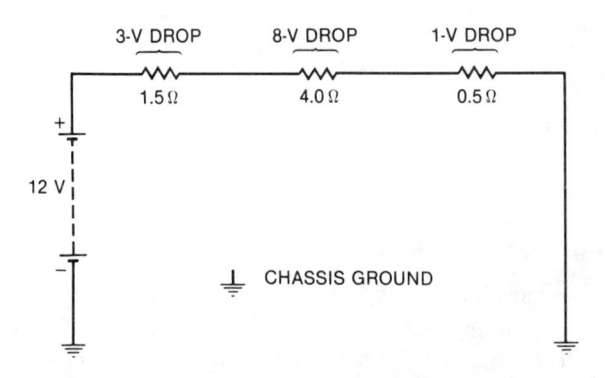

Fig. 1-4. Series circuit showing the voltage drop across each resistance. Note that the sum of the voltage drops must equal the source voltage.

The total resistance of a series circuit is equal to the sum of the individual resistances—in this case, 6 ohms. Knowing the total resistance and source voltage, we can use the formula $I = E/R$ to find the amount of current. Substituting the values, we obtain

$$I = 12 \text{ V}/6 \text{ }\Omega = 2 \text{ A}$$

Next, we use Ohm's law again to determine the voltage drop across each circuit element. Since the same 2 A will flow through each resistance, we can use the formula $E = IR$ to calculate these voltage drops.

$$E = (2 \text{ A})(1.5 \text{ }\Omega) = 3 \text{ V}$$
$$E = (2 \text{ A})(4.0 \text{ }\Omega) = 8 \text{ V}$$
$$E = (2 \text{ A})(0.5 \text{ }\Omega) = \underline{1 \text{ V}}$$
$$12 \text{ V}$$

Introduction to Automotive Solid-State Electronics

Note that these individual voltage drops add up to 12 volts—the same as the source voltage, no more, no less.

By measuring the voltage drops in a circuit, you are in effect measuring the resistance of each element in the circuit. The greater the voltage drop, the greater the resistance. Voltage-drop measurements are a common way of checking circuit resistances in automotive electrical systems. When part of a circuit develops excessive resistance—perhaps due to a bad connection—the portion containing the excessive resistance will show a higher-than-normal voltage drop. This is shown in Fig. 1–5. Normally, automotive wiring is selected to limit voltage drops across the wires to only a few tenths of a volt. Voltage drops in the ground circuit (that portion of the circuit completed by the vehicle chassis and sheet metal) normally do not exceed 0.1 volt.

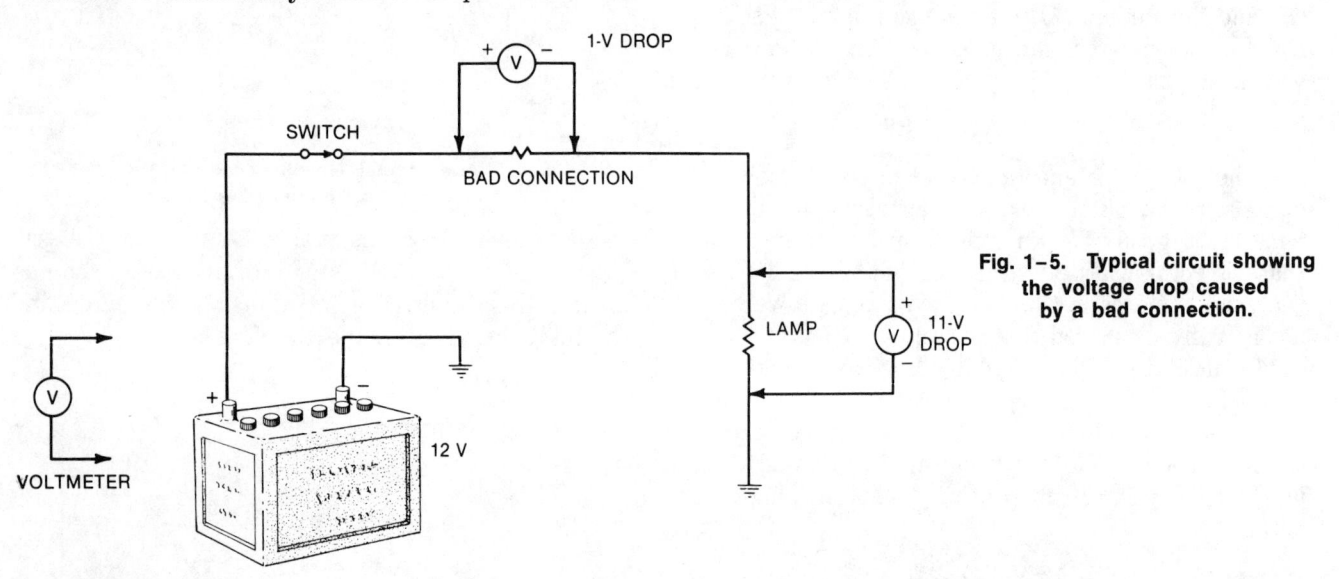

Fig. 1–5. Typical circuit showing the voltage drop caused by a bad connection.

Parallel Circuits

A parallel circuit, unlike a series circuit, contains two or more *branches* (see Fig. 1–6). Each branch can be considered a separate path independent of the others. The total current drawn from the voltage source is the sum of all the currents drawn by each branch. The automotive electrical system as a whole is a good example of parallel circuits.

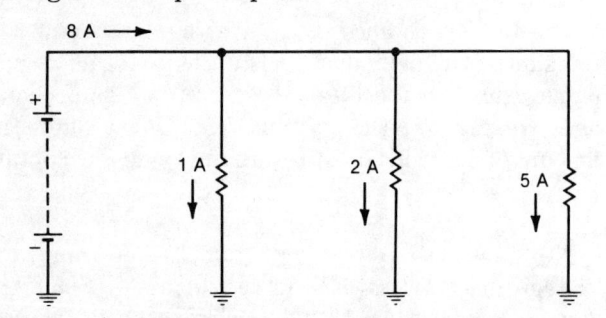

Fig. 1–6. Parallel circuit. The sum of the currents in each branch equals the total current draw.

Each branch of a parallel circuit can be analyzed separately. The individual branches can be either simple circuits, series circuits, or combinations of series and parallel circuits. For example, the headlight circuit is one of the many branches in the complete electrical system. However, each lamp in the headlight circuit is connected in parallel with every other lamp.

Ohm's law applies to parallel circuits, just as it applies to series circuits, by considering each branch by itself. The most important fact to remember about parallel circuits is that the voltage across each branch is the same as the source voltage. The current in any branch is that voltage divided by the resistance of the branch ($I = E/R$).

Series–Parallel Circuits

It is not uncommon for a circuit to consist of elements connected in series and parallel, such as shown in Fig. 1–7. No matter how involved it may seem at first, it can always be resolved into a simple circuit, with some highly complex circuits requiring more resolution than others. For example, the circuit shown in Fig. 1–7 can be reduced to a simple circuit by following the steps shown in Fig. 1–8. A useful formula for accomplishing this is the one that gives the combined resistance of two resistors in parallel:

$$R_{\text{total}} = \frac{R_1 R_2}{R_1 + R_2}$$

Review of Basic Electricity

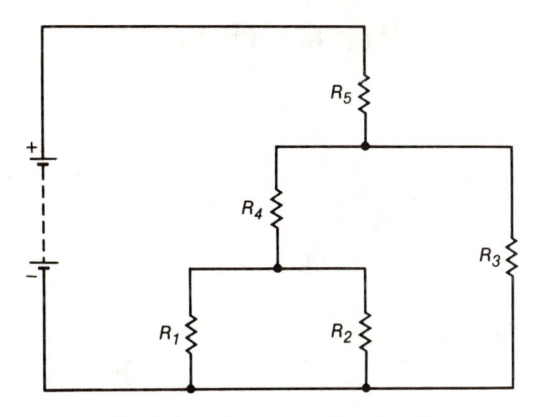

Fig. 1-7. Series-parallel circuit.

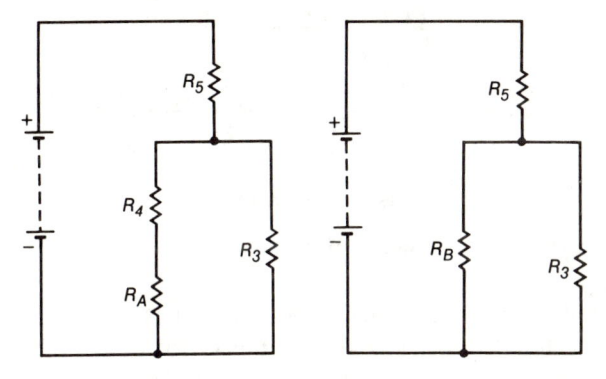

(A.) Combining R_1 and R_2 (R_A). (B.) Combining R_A and R_4 (R_B).

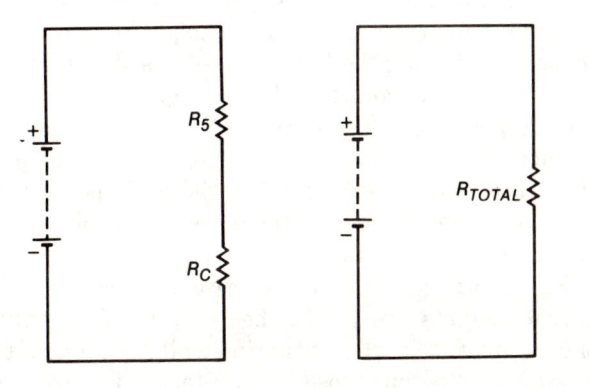

(C.) Combining R_B and R_3 (R_C). (D.) Combining R_C and R_5 (R_{TOTAL}).

Fig. 1-8. Step-by-step resolution of complex circuit to a simple circuit.

Remember that the equivalent resistance of two or more resistors connected in series is just the sum of the individual resistances.

Let us start by resolving resistances R_1 and R_2, connected in parallel, into one equivalent resistance R_A (see Fig. 1-8A). Using the formula given above we find

$$R_A = \frac{R_1 R_2}{R_1 + R_2}$$

Next, resistances R_A and R_4 are seen to be connected in series (Fig. 1-8B) and can be added to give

$$R_B = R_A + R_4$$

In Fig. 1-8C we resolve resistances R_B and R_3, connected in parallel, into a single resistance R_C:

$$R_C = \frac{R_B R_3}{R_B + R_3}$$

In the last step, we find the total equivalent resistance of the complex circuit (see Fig. 1-8D) to be

$$R_{TOTAL} = R_C + R_5$$

since R_C and R_5 in Fig. 1-8C are connected in series.

MAGNETISM

Although man has known about magnetism for thousands of years, there are still some unanswered questions concerning its origin. Magnetism was initially discovered in *loadstone*, a type of natural magnet. At first the stones were just curiosities and thought to possess magical properties, but, during the Middle Ages, the directive properties of loadstone, or needles magnetized by it, were discovered and the magnetic compass was born.

It was not until the 19th century that scientists investigating magnetism learned to create it electrically. From that point on, great advances were made in magnetic theory and its practical applications. Harnessing this strange force has truly been one of man's most notable achievements.

Magnetic Properties of Matter

We normally think of magnetism as affecting only iron or steel (i.e., ferromagnetic materials), but all substances—whether gas, liquid, or solid—are influenced to varying degrees by magnetism. These substances react in one of two ways. If a material is attracted by a magnet, it is said to be *paramagnetic*; if it is repelled by a magnet, it is *diamagnetic*. Iron and other ferromagnetic materials are paramagnetic.

Permanent Magnets

A permanent magnet is either a natural magnet, like loadstone, or an artificial one that retains its magnetism after being magnetized. A magnet always has two poles—a north pole and a south pole. No matter how small a magnet is cut, it will always retain its two poles. The end of a magnet that points to the earth's magnetic north pole is called the north pole of that magnet. Since like poles of a magnet

Introduction to Automotive Solid-State Electronics

repel and unlike poles attract, we are forced to admit that, by definition, the earth's south magnetic pole is actually located in the vicinity of the earth's geographic north pole!

A magnet is surrounded by a magnetic field, shown graphically as lines of force running from the north pole to the south pole of the magnet (see Fig. 1–9). This field can be observed by holding a bar magnet under a horizontal piece of cardboard on which some iron filings have been sprinkled. The filings will line up along the magnetic force lines and reveal the shape of the magnet's field. Magnetic lines of force never cross, although they can be distorted by other lines of magnetic force.

Magnetic strength is determined by the number of lines in the magnetic field, the *flux density*. This is a measure of the number of lines of magnetic force in a given area, similar to the specification of pressure in pounds per square inch.

The permanent magnet finds a number of uses in the modern automobile. Many dashboard instruments are operated electromagnetically and require that a small permanent magnet be attached to the pointer shaft. Most speedometers utilize a magnet and electronic ignition systems frequently use a magnetic pulse generator that incorporates a permanent magnet. Such magnets even find use in oil-pan drain plugs to attract bits of ferrous metal.

ELECTROMAGNETISM

It was discovered in the early 19th century that magnetism is created whenever current flows through a conductor. If current passes through a copper wire, which is nonmagnetic, a magnetic field can be observed around that wire. The effect is shown in Fig. 1–10. The magnetic lines of force are the same as those produced by a permanent magnet and react in the same manner. If two such current-carrying conductors are placed parallel to each other, they will attract or repel one another, depending on the relative direction of the currents. A similar attraction or repulsion would occur if the conductors were placed in the field of a permanent magnet.

The magnetic field about a current-carrying conductor can be increased by wrapping the conductor in the form of a coil. This configuration is an *electromagnet* and finds many applications in automotive systems, such as relays, magnetic switches, solenoids, ignition coils, and alternator field coils. The magnetic strength of such a coil is determined by the number of turns and the current and is expressed in *ampere-turns*. See Fig. 1–10. The greater the number of ampere-turns (amperes times the number of turns), the greater the magnetic strength.

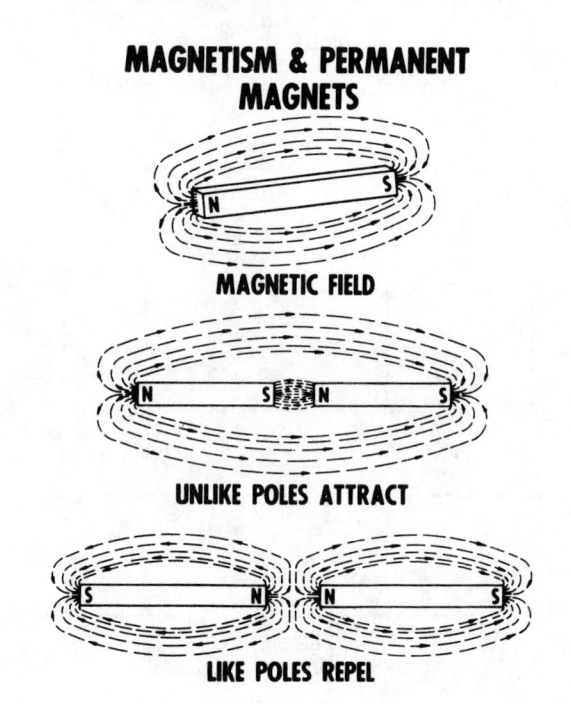

Fig. 1–9. Magnetic field surrounding a bar magnet.

Like a permanent magnet, an electromagnet has north and south poles. The polarity depends on the direction of current through the coil. This can be determined by using the *left-hand rule* as shown in Fig. 1–11. When the coil is held in the left hand with the fingers pointing in the direction of electron flow (remember that electron flow is from negative to positive), the thumb then points to the coil's north pole. As you can see, reversing the direction of coil current reverses the magnetic polarity of the coil.

Fig. 1–12 shows various electrical devices in which electromagnets are used. The relay requires only a small current in the actuating coil to attract the *armature* and thus close the contacts. The contacts can control a much larger current. Thus, the relay is a device whereby a small current can control a much larger current. The magnetic switch, which is actually a heavy-duty relay, is a good example of this. It is used to control the starter-motor current in many vehicles. Although the magnetic switch requires only a few amperes to operate—an amount easily handled by the ignition/starter switch—it can safely carry the several hundred amperes needed by the starter motor.

The solenoid is distinct from the relay and magnetic switch in that it can do a relatively large amount of work. It may or may not operate electrical contacts, depending on the application. The current

Review of Basic Electricity

COMBINED MAGNETIC FIELDS

THE FIELD OF FORCE IS STRENGTHENED BETWEEN TWO PARALLEL CONDUCTORS WHEN CURRENT FLOWS IN OPPOSITE DIRECTIONS. CONDUCTORS ARE PUSHED AWAY FROM EACH OTHER.

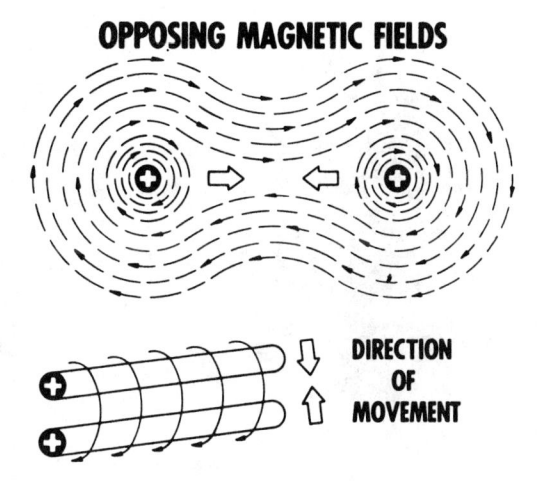

OPPOSING MAGNETIC FIELDS

THE FIELD OF FORCE IS WEAKENED BETWEEN TWO PARALLEL CONDUCTORS WHEN CURRENT FLOWS IN THE SAME DIRECTION. CONDUCTORS ARE DRAWN TOGETHER.

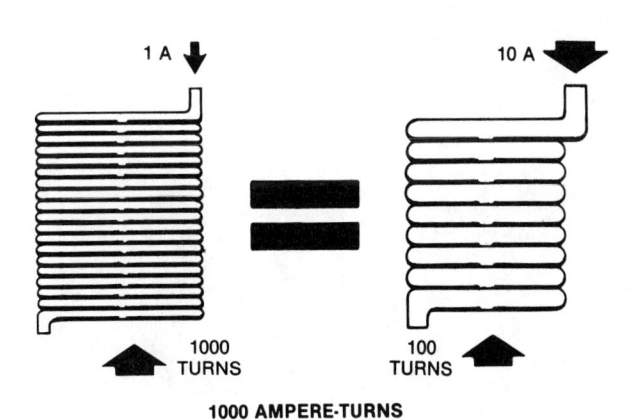

Fig. 1-10. Magnetic field surrounding a current-carrying conductor and the equivalent magnetic strength of two electromagnets.

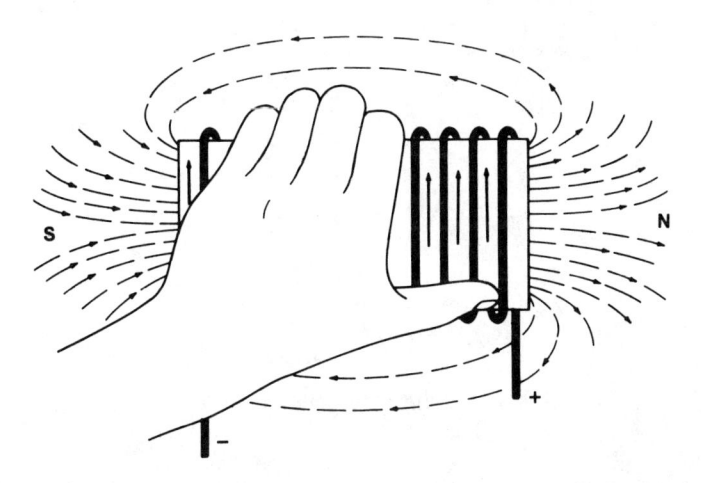

Fig. 1-11. Left-hand rule to determine magnetic polarity.

through the solenoid coil causes a plunger to move in or out and, therefore, provides some mechanical function. When used in emission-control systems, the solenoid may open or close a vacuum passage, advance or retard ignition timing, and change idle speed. In a starter control circuit, it engages the starter drive gear in the flywheel and closes the circuit between the battery and the starter motor. In this last case, the solenoid does double duty.

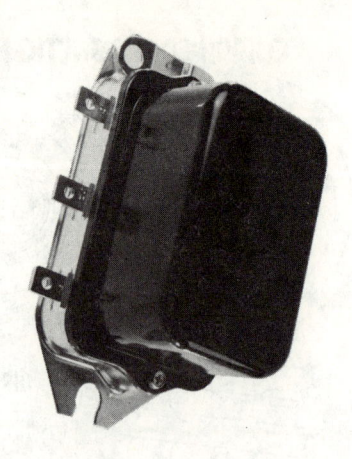

Introduction to Automotive Solid-State Electronics

Courtesy C. E. Niehoff and Company

Fig. 1–12. Typical electrical devices using electromagnets.

CHAPTER 2

Introduction to Solid-State Electronics

In this chapter we explore the basic building blocks of solid-state electronic circuits, principally diodes and transistors. We touch only lightly on the theory of solid-state physics, concentrating primarily on the practical aspects of solid-state applications.

SEMICONDUCTOR MATERIAL

The solid-state devices to be discussed are made from materials known as *semiconductors*. A semiconductor is a substance that is neither a good conductor nor a good insulator—hence, it is a "semi conductor." The principal materials that exhibit these characteristics are the elements silicon and germanium—silicon being the most commonly used material. In practice, the semiconductor material is specially treated to impart to it certain qualities that enhance its function. Thus, it is treated to become either p-type (positive) or n-type (negative) material. Just how these are assembled in an actual device will be covered later.

DIODES

One of the simplest of the solid-state devices is the diode. If you think of it as a one-way valve (remembering the water analogy), you will just about have mastered its function. Although there are a number of different types in the diode group, they all work in basically the same manner. We will start with the most common form—the power diode or rectifier.

Diode Elements

Fig. 2–1 shows the diode symbol. This symbol remains essentially the same for all types of diodes. We will cover the minor variations in this symbol when we discuss the other diode types. The diode, being known as a two-terminal device, has two elements: (1) the anode and (2) the cathode. The anode side is identified by an arrow head and the cathode side by a flat bar. This configuration shows you instantly which way the current will flow through the diode. Using the conventional theory of current flow (i.e., from a positive terminal to a negative terminal), current flows through the diode in the direction of the arrow. Should the diode be reversed, or the direction of current flow be reversed, the diode will block any current flow. This is illustrated in Fig. 2–2.

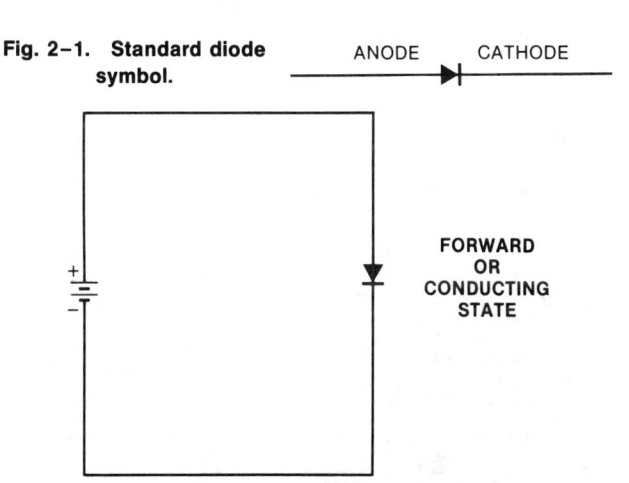

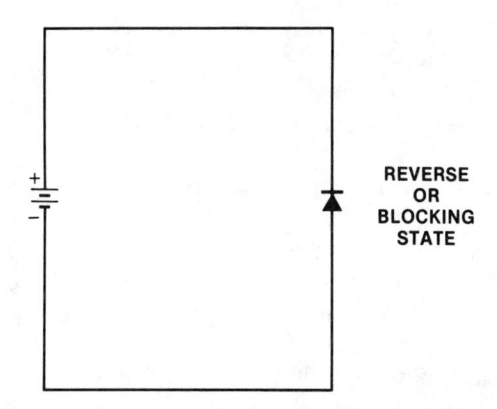

Fig. 2-1. Standard diode symbol.

Fig. 2-2. Diode connected in both its forward and reverse state.

Diode Uses

One of the most common uses of diodes in automotive systems will be found in alternators, where they

Introduction to Automotive Solid-State Electronics

serve a dual function (see Fig. 2-3). First, they serve to rectify the alternating current generated by the stator windings to direct current for charging the battery (more on this in a later chapter). Second, they function as an automatic, solid-state switch to connect the alternator to the battery and other electrical systems. Because of their ability to pass current in one direction and block it in the other, they allow current to flow from the alternator to the battery, but block any reverse current flow from the battery to the alternator. This prevents the battery from discharging through the alternator when the vehicle is not running.

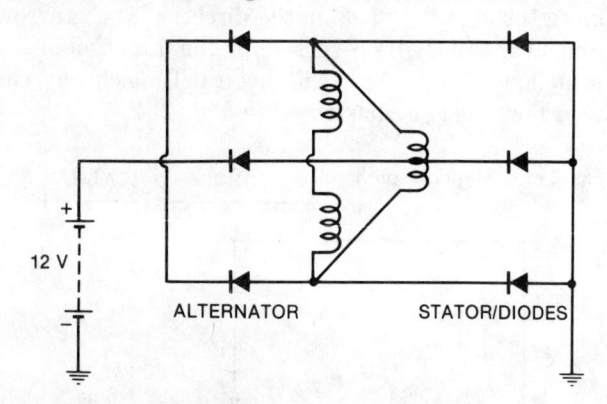

Fig. 2-3. Diodes used in a rectifier assembly for an alternator.

Diode Ratings

In order to fully understand diode operation, you must become familiar with several other aspects of diode performance. Like any other electrical device, diodes have certain ratings that must be observed and should not be exceeded. First, there is the *forward current rating*. This tells you how much current, measured in either amperes or milliamperes, can safely pass through the diode without causing damage or destroying it.

Fig. 2-4 shows what happens as current flows through a diode in the forward direction. There is a voltage drop across the diode, just as there is when current flows through a resistor. However, unlike a resistor, the voltage drop across a diode remains nearly constant regardless of the current. The forward voltage drop in a diode increases only slightly as the current increases. In the case of a silicon diode, which is the only type in widespread use today, the voltage drop holds relatively constant at about 0.7 V.

Heat Dissipation in Diodes—Whenever you have a voltage drop across a device carrying current, you will also have heat. Using the power formula, we can calculate the amount of heat that must be dissipated

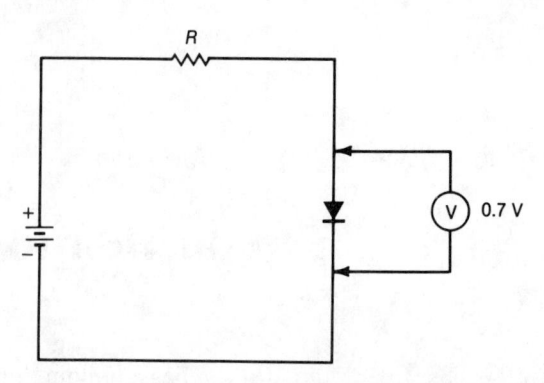

Fig. 2-4. Typical voltage drop across a diode in its forward or conducting state.

from a diode in order to prevent circuit damage. Assume that the diode is passing 10 amperes and that its forward voltage drop is 0.7 volt. Then

$$P = I \times E$$

where,
  P is the power in watts,
  I is the current in amperes,
  E is the forward voltage drop in volts.
  or
$$P = 10 \times 0.7 = 7 \text{ watts}$$

This means that 7 W of heat are being generated in the diode. If this heat is not dissipated fast enough, the diode's internal temperature will rise to the point where the silicon will melt, thus destroying the diode. This will manifest itself as either a shorted or open diode. In either case, the diode function will be lost.

Small diodes, those designed to carry low amounts of current (usually less than one ampere), need no special provision for dissipating this internal heat. As long as the forward current does not exceed the specified rating, internal heating will be held to safe limits. Note, however, that if the diode's surrounding temperature (referred to as the *ambient* temperature) is unusually high, it will take less internal heat to bring the diode temperature up to the melting point. When diodes are operated in high ambient temperature, their forward currents must be "derated" to prevent failure.

Heat Sinks—When diodes are designed to carry higher currents (typically, more than several amperes), they are usually mounted in "heat sinks." A heat sink acts just like a radiator: it draws heat from the device and delivers it to the surrounding air. In effect, it greatly increases the effective surface area of the diode to allow its internal heat to be more rapidly dissipated. Thus a diode that could handle safely only a few amperes by itself, could, with a suitable heat sink, handle 10 amperes or more. The diodes in alternators are typical of heat-sink mounted devices.

Reverse State of Diode

When diodes are operating in the reverse or blocking mode, we assume that no current flows. For practical circuit analysis this is a reasonably safe assumption. However, it is not strictly true. There is always a very slight amount of what is called *reverse leakage current*. In small diodes it is measured in microamperes (millionths of an ampere), while in larger, rectifier-type diodes it is usually measured in milliamperes (thousandths of an ampere).

It is because of this reverse leakage current that diodes also have a specification known as the *peak inverse voltage* rating. Depending on the type of diode, this can be anywhere from 50 to 1000 V or more. It means that the diode can stand no more than its peak inverse voltage rating when it is operating in the blocking mode. If this rating is exceeded, the diode can be destroyed by the generated heat just as it could be destroyed with excessive forward current. Exceeding the inverse voltage rating for even a split second can be enough to cause permanent damage or outright failure.

The discussion about diode ratings and the use of heat sinks also applies to transistors. Most diode and transistor failures are caused by excessive internal heat or voltages in excess of their ratings.

TRANSISTORS

The development of the transistor was perhaps the most significant milestone ever to occur in the electronics industry. Without this invention, first announced by Bell Telephone Laboratories in 1948, most of the electronic devices we now know would never have been possible—at least not in their present form. In essence, the transistor has replaced the vacuum or radio tube (which in itself is not too old, dating only from the early 1900s). In so doing, it has opened up whole new areas where vacuum tubes would have proven too impractical for use.

The ever-present pocket calculator, some no larger than a credit card, would expand to the size of an average living room if it had to be made with vacuum tubes. The advantages of the transistor are its reduced size (literally microscopic), small manufacturing cost (a fraction of a penny when used in integrated circuits), and almost infinite life. Small wonder then that this essentially simple device has spawned such a multitude of electronic products, both useful and trivial.

Transistor Elements

Referring to Fig. 2-5, note that the transistor has some of the elements of the diode just discussed. In fact, we can consider the transistor to be a type of "controllable" diode. In addition to passing or blocking current, the transistor can control the *amount* of current passing through it. Because of this, it can function as (1) an amplifier or (2) a switch. In automotive applications, it is the switching function that is of most interest to us. Let us now consider how the basic transistor works.

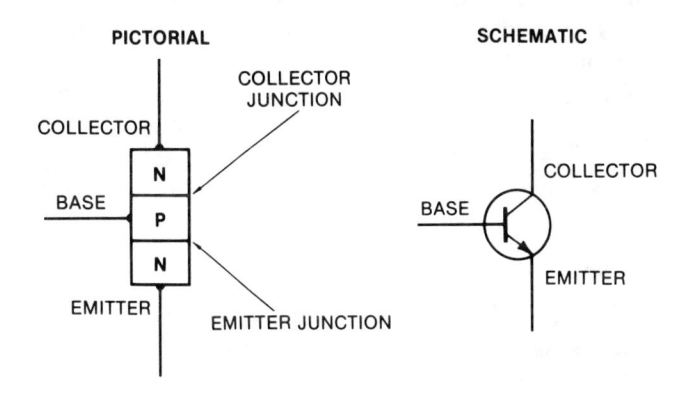

(A) Npn transistor.

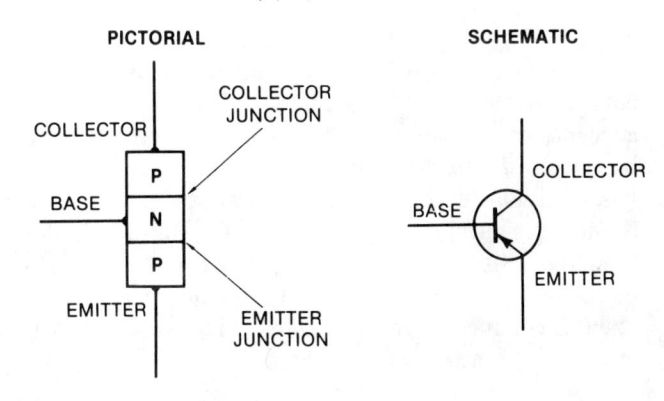

(B) Pnp transistor.

Fig. 2-5. Standard transistor symbols and construction principles for both npn and pnp types.

Construction of the Transistor

Most present-day transistors are made from silicon, the same material used for diodes. (Germanium, a once-popular semiconductor material, is seldom used nowadays.) As shown in Fig. 2-5, there are three elements: (1) the emitter, (2) the collector, and (3) the base. Also, notice that the emitter and base "junction" resemble the symbol for a diode. In fact, it is possible to look upon the transistor as two diodes sharing a common base. This concept is a useful one

Introduction to Automotive Solid-State Electronics

should you ever wish to make a quick check of a transistor with an ohmmeter. (More on this later, though.)

Fig. 2-5 shows the internal construction of a typical transistor. This is merely a schematic representation to show the theory of transistor construction, not necessarily the way an actual transistor is built. The basic material is silicon (or germanium in earlier transistors) to which certain impurities, such as arsenic, antimony, and gallium, have been added in order to form either n-type or p-type material. Slices of these materials are bonded together to form either a p-n-p or an n-p-n "sandwich." The center element in this sandwich becomes the *base;* the other two become the *collector* and *emitter.* Thus two basic types of transistors are formed, the pnp type and the npn type. They function in the same manner, except their polarities are different. Obviously, you cannot substitute one type for the other.

Transistor Function

Fig. 2-6 shows an npn transistor in a typical circuit. Assume that the load represents a lamp whose brightness we wish to control. Note that the main circuit for the lamp current is through the collector and emitter. This current, and hence lamp brightness, will be controlled by adjusting the amount of *base* current. It takes only a small amount of base current to control a large amount of collector-emitter current, so we say that the transistor *amplifies* the base current. In this circuit we control the amount of base current with a potentiometer. As the potentiometer is turned, the base becomes more and more positive. Since the base is p-type material (remember, we are using an npn-type transistor), it will start to conduct current. It takes only a small amount of base current to cause a large amount of collector-emitter current to flow, so the lamp turns on and becomes brighter and brighter as the base current increases.

The ratio of the amount of base current required to produce a given amount of collector-emitter current is called the *gain* of the transistor and is expressed by the term "beta." Thus, a transistor with a beta of 100 can cause a one ampere change in collector-emitter current with only a 10 milliampere change in base current. Transistors are designed with different gains for different applications. A low-gain transistor may have a beta of 20, whereas a high-gain device may have a beta in excess of 200.

Practical Transistor Uses

The example shown in Fig. 2-6 is not too practical, except for explanation purposes, since we could have controlled the lamp directly with the potentiometer. But let's see if we couldn't make the lamp turn on automatically when it gets dark and off again when it gets light. Fig. 2-7 shows how this might be accomplished. In place of the potentiometer, we substitute a *voltage divider* consisting of a fixed resistor and a cadmium sulfide (CdS) photocell. A cadmium sulfide photocell is actually a light-sensitive resistor. When no light falls on it, its resistance is quite large; when light hits it, its resistance drops to a low value.

Here is how our automatic lamp controller works. When the photocell is exposed to the dark, its resistance becomes very large. In effect, this places the base of the transistor closer to the positive voltage source and the base current flows. This, in turn, increases the collector-emitter current and the lamp turns on. The circuit responds in the same manner as that of Fig. 2-6 when the potentiometer was turned up.

Fig. 2-6. Npn transistor used in a lamp-control circuit.

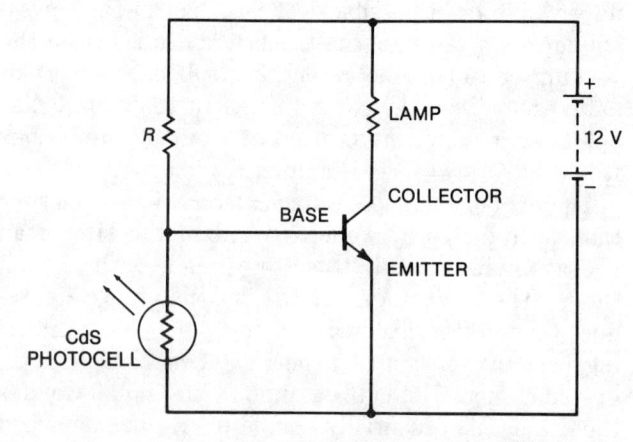

Fig. 2-7. Circuit of Fig. 2-6 being controlled by a CdS photocell.

Introduction to Solid-State Electronics

When daylight strikes the photocell, its resistance drops to a low value. This, in effect, pulls the transistor base away from the positive potential and puts it closer (electrically) to the negative potential, thus reducing or cutting off any base current. Again, it is as though we had turned the potentiometer of Fig. 2-6 almost all the way down. As a result, collector-emitter current is cut off and the lamp extinguishes.

Obviously, the light from the lamp must never be allowed to strike the photocell. If it did, its own light would be signaling the photocell to turn itself off. And when the lamp went off, the photocell would immediately turn it back on again, and so on in a mad, on-off race. In reality, the lamp would be glowing at some brightness between off and on. This situation is known as *positive feedback* and, while it may be undesirable in an automatic lamp system, it is quite useful in other applications. It is used to good advantage in the voltage regulator circuit of an alternator, a system we will cover later.

It should be noted that we could have used a pnp-type transistor for the circuit shown in Fig. 2-7. However, we would have had to reverse the polarity of the voltage source. Other than this, the circuit would have operated in the same manner. This fact gives circuit designers considerable flexibility in designing circuits to meet specific requirements.

In practice, the collector terminal is reverse-biased and the emitter terminal is forward-biased. Thus, an npn transistor would have its collector connected to the positive side of the circuit and its emitter connected to the negative side. The base, because it is p-type material, would have to receive its base current from a positive source. In the case of a pnp transistor, the above polarities would have to be reversed for the transistor to function.

Transistor Ratings

Much of what we discussed about ratings for diodes applies also to transistors, since they are constructed of the same materials. When transistors are required to handle relatively high currents, such as they would in voltage regulators or ignition systems, they are generally mounted on heat sinks in the same manner as diodes. Likewise, they can also be damaged or destroyed if their voltage ratings are exceeded.

Transistor States

Many of the transistor applications in automotive circuits require them to operate as a switch. That is, they must be either turned on (conducting collector-emitter current) or turned off (no collector-emitter current). This is controlled by the amount of base current. When the base current is zero, the transistor is said to be in *cutoff*. Sometimes the base is actually reverse-biased to ensure complete cutoff.

As the base current increases, collector-emitter current also increases, but only up to the maximum dictated by external circuit resistance. If more base current is supplied after this point is reached, the transistor is said to be in *saturation*. The collector current remains at its maximum level. Therefore, to operate as an electronic switch, the transistor must swing from a cutoff state (equivalent to an open switch) to a saturated state (equivalent to a closed switch). This is controlled by the base current which must change quickly from zero to the maximum value needed for saturation. Depending on various factors, a transistor can turn on or off in less than one microsecond (one millionth of a second).

Transistor Check Using Ohmmeter

We mentioned earlier that a transistor can be considered as two diodes sharing a common terminal (the base). For a quick check, remove the transistor from the circuit. Measure the forward and the reverse resistance first between the base-emitter terminals and then between the base-collector terminals. To do this, simply reverse the ohmmeter leads between measurements. The forward resistance should be small while the reverse resistance should be large. Compare your reading with those from a known good transistor of the same type. As a final check, measure the forward and reverse resistance between the collector and emitter terminals.

CHAPTER 3

Charging Systems

Although each electrical system in a vehicle has its own specific function, none would be possible without the support of the charging system. The charging system is to the automobile what a power line is to our home. Things would be rather primitive without either one. And as the automobile becomes more and more "electrified," the task and the burden of the charging system become even greater.

Stated briefly, the purpose of the charging system is to replenish the power drawn from the battery during cranking and to supply operating current to all electrical accessories during normal driving. True, a vehicle can operate for a short period, depending on the reserve capacity of the battery, in the event of charging failure. But even this emergency mode depends on how well the system was functioning before it failed. As often happens, the battery is never properly charged just before total failure occurs, and consequently there is no reserve capacity to supply the necessary electrical systems.

In this chapter, we will review the principles of charging-system operation, the types of charging systems and how they work, and, in particular, the regulators that control them.

BATTERY CONSTRUCTION

Since a battery is a chemical device, we must consider its chemical makeup to understand just how it works. There are literally hundreds of ways to make batteries, but almost all of them follow the same format: two dissimilar conductive materials separated by an electrolyte. For instance, a piece of newspaper wetted with vinegar (the electrolyte) sandwiched between a copper penny and an iron plate will produce a feeble current. Technically, this is a battery, although impractical for most applications since it lacks "capacity." In a battery for automotive application, capacity is vitally important.

Automotive batteries are classed as lead-acid storage batteries, which describes quite well their construction and operation. Usually these consist of three or six cells connected in series to form either a 6- or a 12-volt battery. Each cell can be considered a separate battery. The plates of these cells are made of lead (pure sponge lead for the negative plates and lead peroxide for the positive plates) and immersed in an electrolyte of dilute sulfuric acid—hence the term lead-acid. The word "storage" means that it can operate in either a charging or discharging mode. This distinguishes it from the so-called "primary" or nonstorage types of batteries used in flashlights and similar devices.

Charging and Discharging Processes

A lead-acid battery produces current through the chemical reaction shown in Fig. 3–1. This reaction takes place whenever the two plates (or two groups of plates) are connected together through a current-carrying conductor. Here is what happens. The sulfuric acid (H_2SO_4) in the electrolyte wants to combine with the lead (Pb) in the negative plate to form lead sulfate ($PbSO_4$). But to do this, the lead must *give up* electrons. Meanwhile, the sulfuric acid also wants to combine with the lead peroxide (PbO_2) in the positive plate to form lead sulfate. But to do this, it *needs* electrons. The current-carrying conductor connected between the plates provides for the necessary transfer or flow of electrons and the reaction takes place. This flow of electrons *is* the current produced by the battery.

$$Pb + PbO_2 + 4H_3O^+ + 2SO_4^{2-} \underset{\text{DISCHARGE}}{\overset{\text{CHARGE}}{\rightleftarrows}} 2PbSO_4 + 6H_2O$$

Fig. 3–1. Basic chemical reaction of a battery during charge and discharge.

The above reaction can be reversed by forcing the electrons to flow in the opposite direction. This is what takes place during charging. When this happens, the lead sulfate that formed on the positive and negative plates during discharge gets converted back into sulfuric acid and the battery becomes recharged.

Since the action of a battery during charge and dis-

Introduction to Automotive Solid-State Electronics

charge is so important, let's summarize what we have covered. When a battery discharges, the acid "leaves" the electrolyte and combines with the plate material to form lead sulfate. If the battery were to be thoroughly discharged (which seldom happens), the electrolyte would be reduced to pure water. When a battery charges, the lead sulfate on the plates combines with the water to produce a sulfuric acid solution and the plates are restored to their original state.

Battery Failure

From this description, it would seem that a battery could go on forever, charging and discharging, and never wear out. Practical experience tells us otherwise. Two common causes of battery failure are loss of active material from the positive plates (cycling failure) and sulfation. Cycling failure is brought on by the charge–discharge cycle. Each time this occurs, a small amount of the lead peroxide in the grid of the positive plate gets pushed out and falls to the bottom of the cell chamber. As a result, the battery gradually loses capacity (its ability to produce high current) until it can no longer provide cranking power.

Sulfation has a similar effect (i.e., loss of capacity) but is caused by a different condition. The term "sulfation" applies to the lead sulfate that forms normally on the plates during discharge. As long as this is quickly converted back to acid by the charging process, no damage occurs. But when the lead sulfate remains on the plates for any length of time, due to lack of proper charging, it tends to harden. The longer it remains, the harder it gets and the more resistant it becomes to any subsequent charging current. It converts back to acid, if at all, only with difficulty. Since lead sulfate is not an active material (it has actually replaced the active material), the result is a loss of battery capacity.

Winter driving tends to promote sulfation. Electrical loads are high, driving speeds are slow, and battery temperatures are low (cold batteries are much harder to charge than warm ones). The task of keeping the battery fully charged in winter is considerably greater than at other times so it is especially important that the charging system be thoroughly checked for proper operating limits prior to the onset of winter. However, this does not mean that the system can be neglected during other seasons.

A rather insidious type of battery problem can actually develop during warm-weather driving and does not show up until the heavy starting demands of winter. This is overcharging, due to a higher-than-normal regulator setting. The condition leads to premature failure of the positive plates of the battery. But since warm-weather battery demands are light, the fault goes unnoticed—until some bitterly cold, wintry morning. Most people will assume that the problem was caused, or brought on, by the cold weather but, in this case, it was the good old summertime (combined with an improper regulator setting) that did the battery in.

As you can see, there is a very close relationship between the battery and the charging system. In fact, you could go so far as to say that the very life span of the battery is, in large measure, controlled by the charging system. It is not enough that the system merely be operating; it must be operating within prescribed limits.

ALTERNATORS

Until a few years ago, generators were the major element in the charging system. Today it is the alternator. In construction, the two devices are quite similar; in principle of operation, they are almost identical. Since alternators constitute the majority of automotive applications, we will concentrate on this particular unit.

Basic Alternator

An alternator (or generator) works on the principle of *electromagnetic induction*. This rather fearsome term reduces down to three simple elements: (1) a current-carrying conductor, (2) a magnetic field, and (3) motion. Let's take the simplest case as shown in Fig. 3-2. The magnetic field is provided by an old-fashioned horseshoe magnet. The current-carrying conductor is merely a piece of wire. And for motion, we will simply move the wire in and out of the horseshoe magnet.

To prove that we have indeed produced electricity, we connect a voltmeter to the ends of our current-carrying wire. The direction of motion of the wire is very important: it must *cut across* the magnetic lines of force. These are invisible lines of force that are assumed to flow from the north pole of a magnet to the south pole. The more lines there are, the stronger the magnetic field or, technically speaking, the greater the *magnetic flux*. If we move the wire (the current-carrying conductor) *parallel* to these lines of force, nothing happens. But if we move it *across* the lines of force, we will see the voltmeter deflect, indicating that a current has been induced.

Note also that if the wire is moved in one direction, the meter deflects one way and if the wire is moved in the opposite direction, the meter deflects the op-

Charging Systems

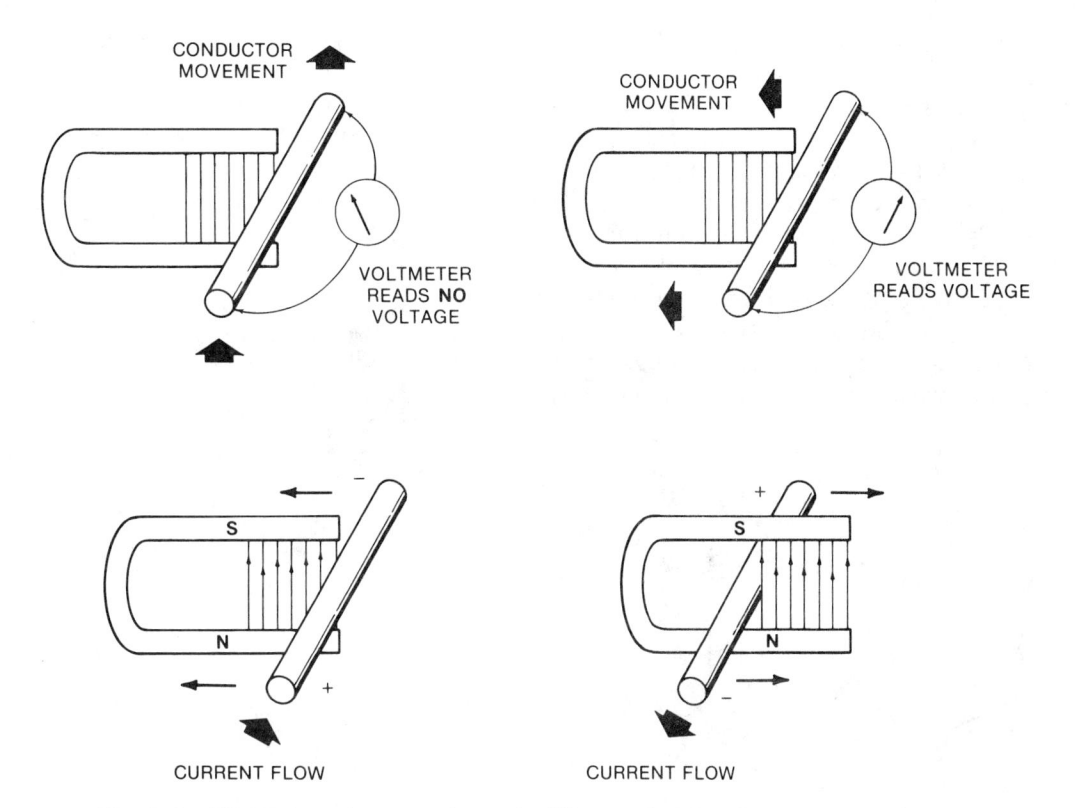

Fig. 3-2. Effect of moving a conductor in different directions in a magnetic field.

posite way. From this observation, we can surmise that the direction of the induced current depends on the direction of motion of the current-carrying conductor with respect to the magnetic field or, conversely, on the polarity of the magnetic field with respect to the direction of motion of the conductor. In other words, if we reversed the polarity of the horseshoe magnet by turning it upside down, the meter would have deflected in the opposite direction to what it did originally.

Now let's try a simple variation. Instead of moving the *conductor* through the magnetic lines of force, let's hold the conductor stationary and move the *magnet*. As indicated by the deflection of the meter, we get exactly the same results. This proves that the necessary motion between the conductor and the magnetic field is only relative; it makes no difference which moves, as long as one does and the other does not. This is exactly what takes place in an alternator.

In an alternator, the current-carrying conductors (called the *stator windings*) remain stationary and the magnetic field (called the *rotor*) revolves within the conductors. (In a generator the reverse is true; the conductors revolve and the field remains fixed.) A basic alternator is shown in Fig. 3-3. The important point to note here is that the polarity of the induced current changes as the magnetic field revolves. The reason, of course, is that the polarity of the rotating magnet, with respect to a given conductor, reverses itself every half revolution. In reality, this simple alternator is generating an alternating current (hence the term "alternator").

Improvement of the Basic Alternator

An alternating current may be satisfactory for some applications, such as lighting a lamp, but it is totally unsatisfactory for automotive applications. We can correct this deficiency quite easily and will do so shortly. But for now, let us see if we can improve upon our basic alternator.

The alternator's major shortcoming is that it is rather inefficient. In its present form it would be called a single-phase alternator. This can be seen better by referring to Fig. 3-3. As the magnet rotates through one complete revolution, the induced output current goes through one complete cycle or phase. It goes from zero volts to a maximum positive voltage, back to zero and then up to a maximum negative voltage, and finally back to zero where the whole process starts over again. We can increase the efficiency by adding two additional sets of stator windings, staggering them so that each is 120° of rotation from the other. This makes it a three-phase alternator.

We can further increase the efficiency by adding

23

Introduction to Automotive Solid-State Electronics

more poles to the rotor. This way, we can get a number of cycles of induced current in each stator winding for each revolution of the rotor. How we get a multipole rotor is shown in Fig. 3–4. This particular rotor has a total of 14 poles, formed by two seven-fingered "cups" between which is located the electromagnet providing the magnetic field. Since the rotor has in effect seven magnets, each of the stator windings must have seven sets of coils, all connected in series. Fig. 3–5 shows how these coils are arranged on the stator frame. The windings of the other two phases would be similarly dispersed around the stator.

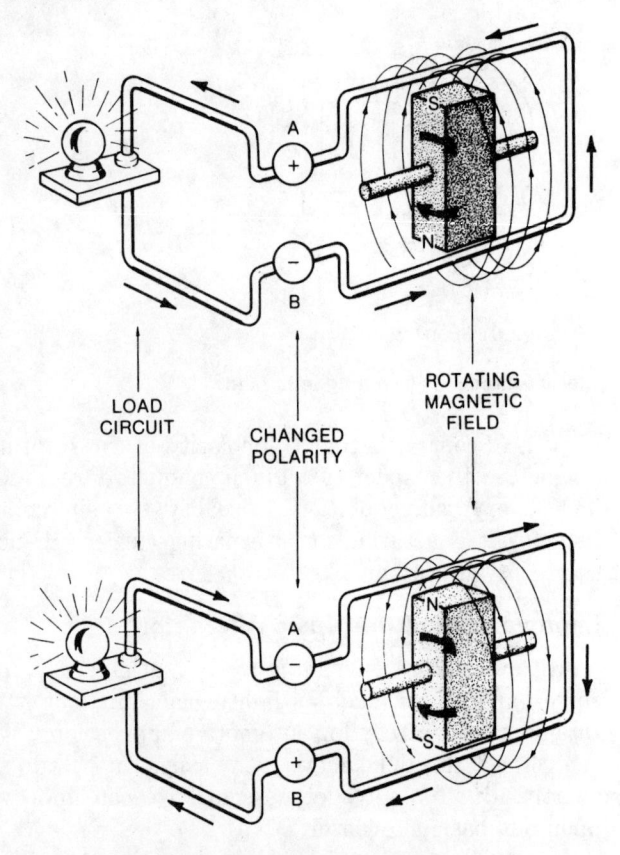

Fig. 3–3. Basic alternator.

Windings—The three phases constitute three sets of windings which must ultimately be connected together. There are two ways this can be accomplished (Fig. 3–6). The first is the *wye* connection (Fig. 3–6A), so-named because of its resemblance to the letter Y. The junction at the center of the Y is sometimes referred to as the "neutral" and occasionally will have a wire leading from it. In these instances, it generally serves to operate a field relay in the regulator.

The second method of stator connection is called the *delta* (Fig. 3–6B) because of its resemblance to

the Greek letter of the same name. Although there are technical reasons for selecting either a delta or a wye configuration (of interest only to the designer), from a practical standpoint it makes no difference which is used. For our purposes, there is no difference in operation, servicing, or testing.

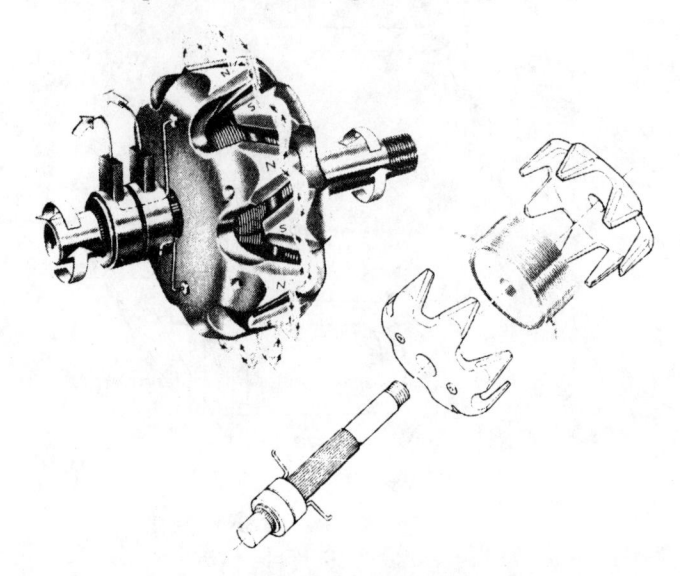

Fig. 3–4. Typical rotor assembly for an alternator.

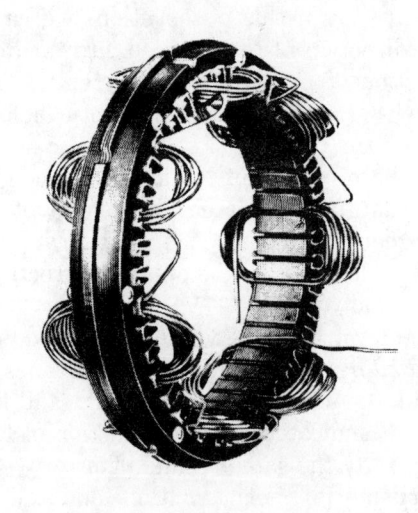

Fig. 3–5. Stator assembly showing the windings installed for only one phase.

At this point it may seem that the simple alternator we started with has become a rather involved affair. We normally think of an electrical circuit as having only two conductors—one going out and another coming back. Now, it seems, we have three conductors. And on top of that they are producing alternating current, when what we really want is di-

CHARGING SYSTEMS

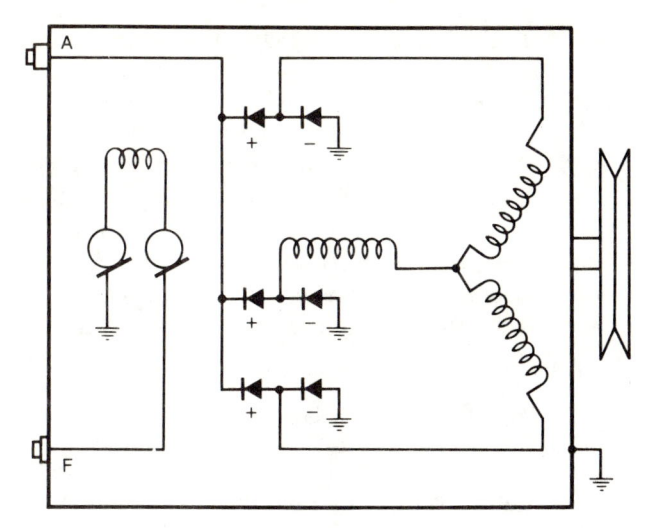

(A) Wye connected.

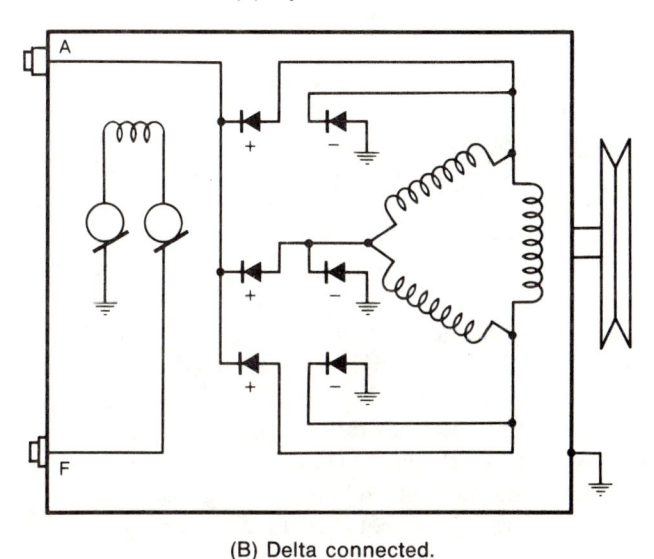

(B) Delta connected.

Fig. 3-6. Two ways in which stator windings can be connected.

rect current. This dilemma is solved quite neatly with the *rectifier assembly* or, as it is also called, the *diode assembly*. Let's review quickly the function of a diode.

Stated briefly, a diode is an electrical one-way valve. It allows current to flow freely in one direction only; i.e., it is *polarity sensitive*. When it is conducting, a diode is said to be forward biased; when it is blocking current, it is said to be reverse biased. These two states are shown in Fig. 3-7. For a fuller discussion of diode operation, see Chapter 2. For now, just remember that a diode passes current in only one direction. Thus the current that passes through is unidirectional or "direct current." In essence, it converts or "rectifies" alternating (bidirectional) current into direct current.

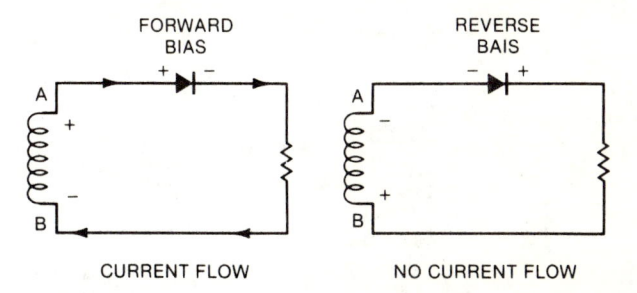

Fig. 3-7. Forward- and reverse-bias connections of a diode.

Although a single diode can rectify an alternating current into direct current, it does not do so as efficiently as we would like. It provides only "half-wave" rectification, as shown schematically in Fig. 3-8. We get direct current only half of the time; the other half of the time we get nothing. What we actually need is "full-wave" rectification which allows us to take advantage of both halves of the alternating input current. This requires a system of four diodes as shown in Fig. 3-9. Now, regardless of the polarity of the incoming alternating current, we always get a unidirectional or direct current coming out of the rectifier. You can prove this to yourself by tracing the current path through the rectifier diodes for any input polarity. Remember, we assume that conventional current flows in the direction of the diode arrow.

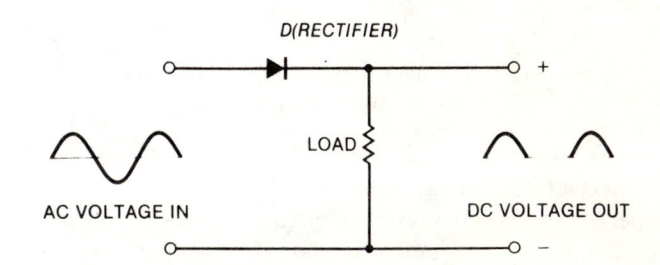

Fig. 3-8. Single diode and half-wave rectification.

Our problem is that we have a three-phase alternator and, therefore, need a three-phase, full-wave rectifier. We solve this simply by taking the above full-wave rectifier and adding two more diodes. This is connected to the stator windings as shown in Fig. 3-10. Because of the way it is connected, each phase "sees" a full-wave rectifier, and, because of the overlapping of the three phases, the rectified output is a remarkably smooth and steady direct current. Notice that although there are three inputs to the rectifier, there are only two outputs—a positive and a negative.

Rotor—Now let's turn our attention to the controlling element in the alternator: the rotor or field

25

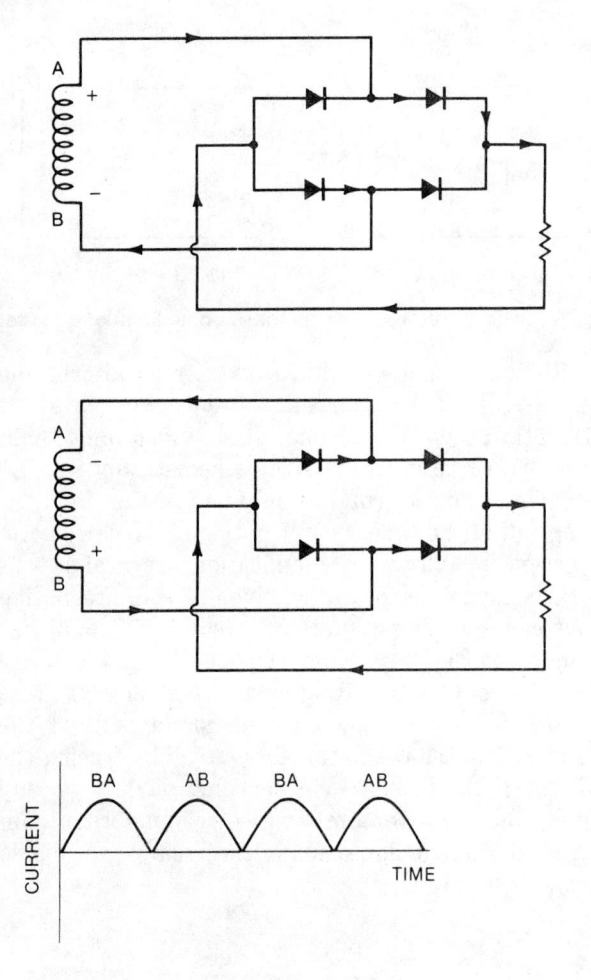

Introduction to Automotive Solid-State Electronics

Fig. 3-9. Four diodes connected as full-wave bridge-type rectifier.

circuit. This is the device by which we control the output of the alternator, or more specifically, the operating voltage of the alternator. Of the three factors needed to generate electricity (conductors, motion, and magnetism), only magnetism provides an effective method of alternator output control. Obviously, we cannot vary the number of conductors in the stator, nor the speed of the rotor (it is continuously varying anyway). But magnetism is easy to control. Since the field coil, which is sandwiched between the finger cups of the rotor (see Fig. 3-4) is nothing more than an electromagnet, all that has to be done to control magnetism is to control the field current. We do this through the voltage regulator.

VOLTAGE REGULATOR

The voltage regulator is the only active control unit in an alternator-driven charging system. A few systems may employ an additional relay, in addition to the voltage regulator, for the purpose of initially energizing the field circuit, but it plays no role in the control of the alternator. Most systems, especially those controlled by solid-state devices, use only the voltage regulator.

Those who are familiar with the older generator-driven charging systems may wonder why two additional controls—the current limiter and the cutout relay—have been omitted. The answer is simply that alternator systems do not need them. The purpose of the cutout relay was to disconnect the generator from the battery, when the engine was stopped, to prevent the battery from discharging back through the generator. The alternator's rectifier or diode assembly accomplishes the same thing electronically, allowing alternator current to flow into the battery but blocking battery current from discharging back through the alternator.

A current limiter was needed in a generator system to prevent its output current from going so high

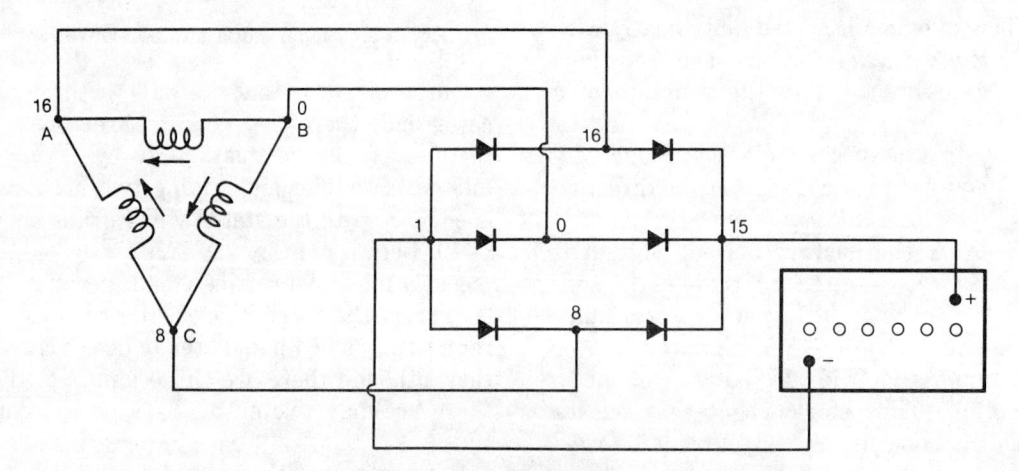

Fig. 3-10. Six diodes connected for full-wave rectification in a three-phase alternator.

Charging Systems

that it would self-destruct the generator. An alternator is self-limiting on output current. A given alternator is designed to produce a maximum of so much output current (say 60 A) and no more, regardless of the demands from the battery and electrical system. Because of this, there is no need for a current limiter.

This brings us to the voltage regulator itself. The purpose of the voltage regulator is to put an upper limit on the voltage that the alternator can produce. As soon as the system voltage tries to rise above this limit, the regulator comes into play and reduces the alternator's output. Two types are currently used to accomplish this regulation: (1) the electromechanical regulator and (2) the solid-state regulator. Both systems work on essentially the same principle, but obviously the circuit details differ. Since the electromechanical method is the older, we will start with it.

Electromechanical Voltage Regulator

Fig. 3–11 shows the basic elements of an electromechanical voltage regulator. This unit is also called a "vibrating voltage regulator," which describes the operating mode quite well. As you can see, the current that flows to the field winding in the rotor (remember, we control output by controlling field current) passes through a set of contacts in the voltage-regulator relay. As long as these contacts remain closed, full field current, and thus full alternator output, occurs.

Notice that the actuating winding under the voltage-regulator relay is connected, in effect, directly across the battery. When the system voltage rises to some predetermined value, the magnetic pull of the actuating winding becomes strong enough to attract the armature of the relay and pull it down. When this happens, the contacts open and field current must then flow through a resistor, thus weakening the field current.

The weakened field current reduces the magnetic strength of the rotor and alternator output falls off. This causes a drop in system voltage resulting in a weakening of the magnetic pull on the relay armature. The relay contacts close and full field current is again allowed to flow. Alternator output increases and system voltage rises and the whole process repeats itself. The voltage-regulator relay contacts open and close rapidly, thereby preventing system voltage from exceeding a preset level. The rapid action of the contacts gives rise to the expression "vibrating voltage regulator."

The setting of the voltage limit is controlled by the spring tension on the voltage-regulator relay armature. See Fig. 3–12. Increasing spring tension raises the regulated voltage while decreasing the tension lowers the voltage. In effect, the spring is counterbalancing the magnetic pull of the actuating coil

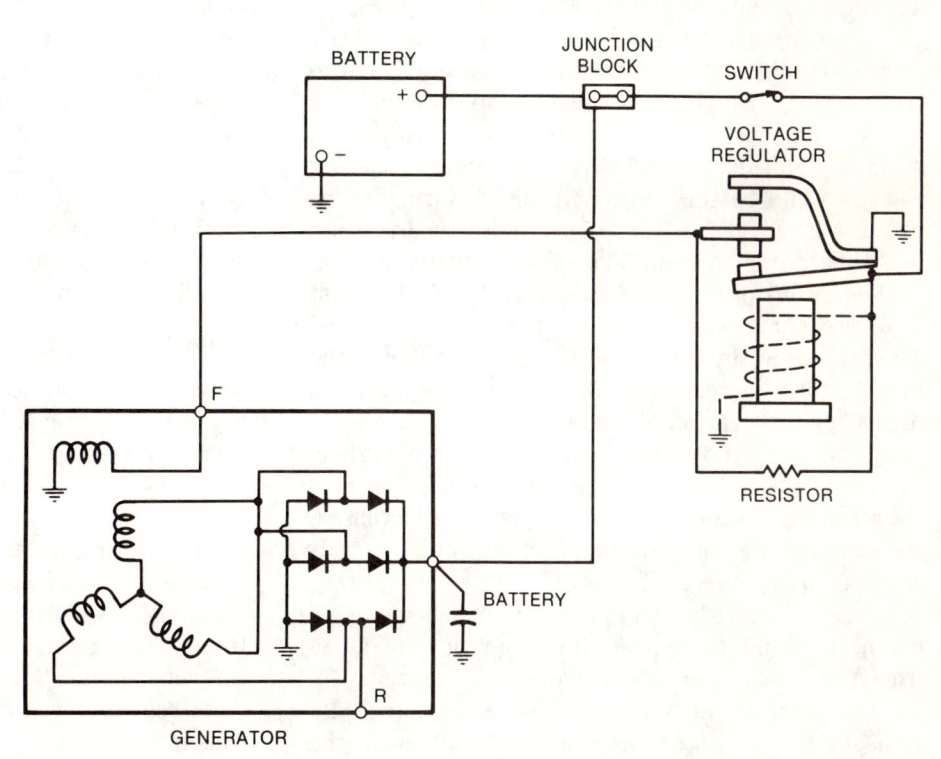

Fig. 3–11. Basic elements of an electromechanical regulator.

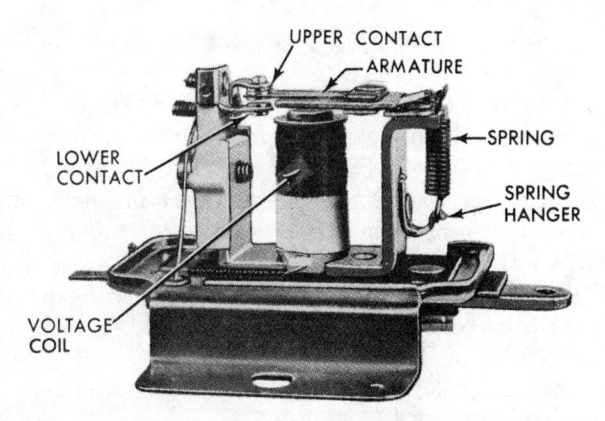

Introduction to Automotive Solid-State Electronics

Fig. 3–12. Typical vibrating voltage regulator. Note the voltage-adjusting spring attached to the armature.

under the relay armature. Of course, the amount of pull exerted by this coil depends on the system voltage and the resistance of this coil. The second factor presents a problem in maintaining a constant or steady level of regulated voltage because the resistance of the coil varies with temperature. When it is cold, its resistance is low and consequently it will develop sufficient pull to overcome the counterforce of the spring at a lower-than-normal system voltage. In other words, the regulated voltage will be lower than desired. On the other hand, when the coil is hot, its resistance is high and therefore the relay operates at a higher-than-normal system voltage. (See ampere-turns in Chapter 1.)

The ideal situation is to have the regulator operate at a constant level of voltage over a wide temperature range. It is even desirable to have the regulated voltage somewhat higher during the warmup period, especially during cold weather, in order to quickly replenish the battery current drained during starting. For this reason, the regulator is usually "temperature compensated." This is generally accomplished in one of two ways (see Fig. 3–13). The first is to use a bimetallic hinge on the relay armature. The hinge is so designed that it becomes stiffer as the temperature decreases. This has the same effect as increasing the tension of the control spring. As a result, the system voltage is regulated at a higher-than-normal level until the regulator reaches a normalized temperature. As the temperature increases, the hinge relaxes its tension and the regulated voltage returns to its specified level.

A second method uses a temperature-sensitive magnetic shunt to bypass the magnetic flux away from the relay armature during cold operation. As the temperature increases, the shunting effect decreases. The effect on the regulated voltage is the

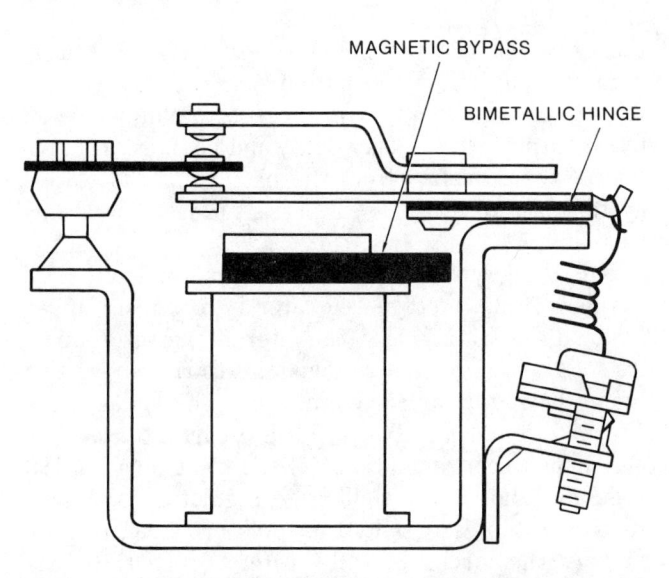

Fig. 3–13. Two methods of achieving temperature compensation in a vibrating voltage regulator.

same as that described above. Because of these temperature-compensating devices, it is important to remember that a charging system should always be tested at its normal operating temperature. This is especially true when comparing the voltage-regulator setting against the specified limits. Normal temperature is assumed to be reached after approximately 10–15 minutes of vehicle operation.

Occasionally a system will employ an additional relay in the regulator. Such relays are frequently called *field relays* and are used to connect the field circuit to the battery (or other "hot" point in the electrical system) when the engine starts. See Fig. 3–14. In some systems they are actuated when the ignition switch is turned on, while in other systems they are actuated by the voltage developed in the neutral wire coming from the alternator. The field relays play no part in the actual regulation of the system voltage; they simply trip on when the engine starts and trip off when it stops.

Solid-State Regulator

Solid-state circuitry lends itself ideally to voltage regulation, especially in automotive charging systems. Among its advantages are (1) low manufacturing costs, (2) small size, and (3) theoretically unlimited life. This last attribute, as experience shows us, is more of a goal than a reality. However, a solid-state regulator does exhibit much greater longevity than does its electromechanical counterpart. The latter is always subject to a finite life span because of contact erosion. Since it can be made small, a solid-state regulator can be installed within an alternator, thus making a one-piece or integral charging system.

Charging Systems

the reverse current but allows it ready passage. The diode is said to be in its *avalanche mode*, shown graphically in Fig. 3-15. Although a standard diode would be seriously damaged or destroyed during such a breakdown, the special construction of the zener's junction prevents this. In fact, this is the exact condition under which a zener must operate if it is to perform the function of voltage regulation.

This breakdown or avalanche point is voltage controlled and fixed for any given zener. However, zeners can be made to exhibit any zener voltage (as the breakdown point is called) from a low of several volts to a high of several hundred volts or more. A zener used in a typical automotive voltage regulator may have a zener voltage of approximately 5 volts.

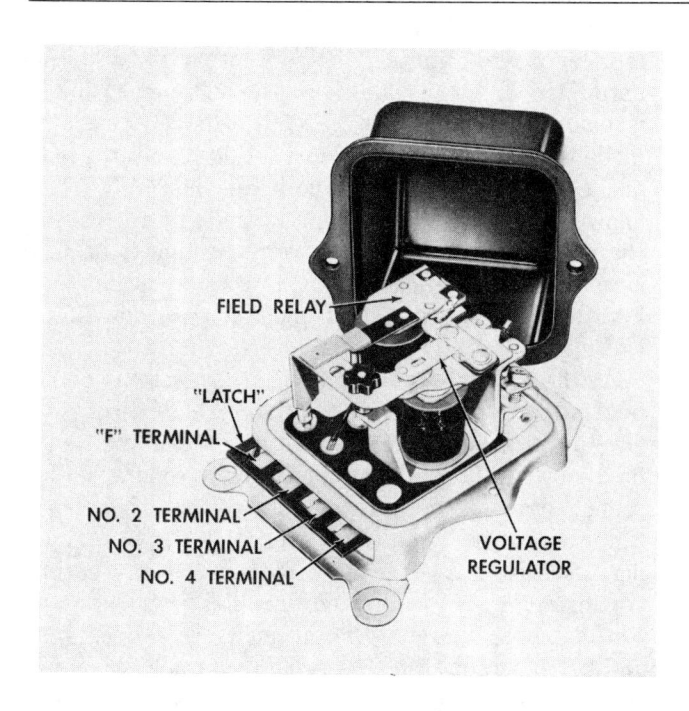

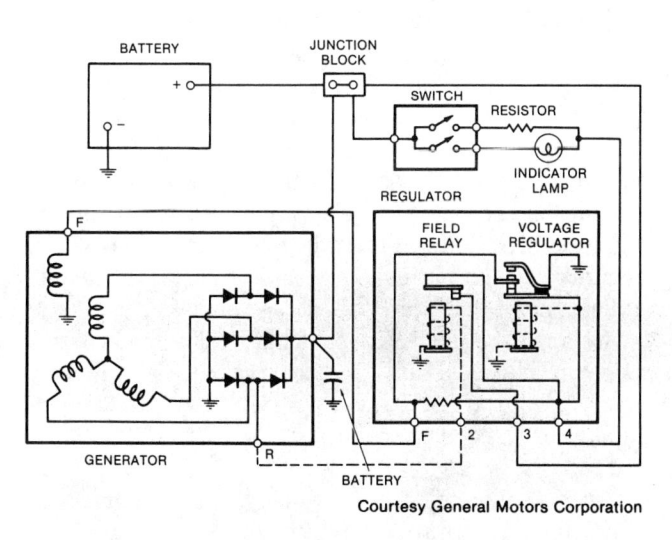

Fig. 3-14. Alternator system employing a field relay.

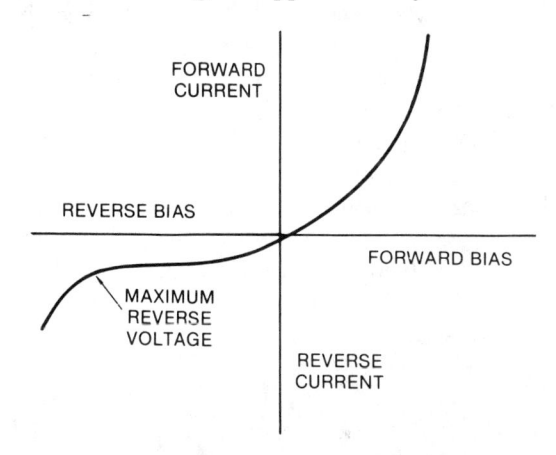

Fig. 3-15. Voltage-current diagram for a diode in avalanche mode.

Let's see how a zener might be used in a very simple (nonautomotive) type of regulator. As shown in Fig. 3-16, all it takes is a zener and a current-limiting resistor. The unregulated voltage enters the circuit from the left. As soon as the unregulated input voltage rises to the zener breakdown voltage, the zener conducts and clamps the output voltage at the zener level. Should the unregulated input voltage rise above the zener voltage, the excess voltage (the difference between the unregulated voltage and the zener voltage) will be absorbed by the current-

The heart of a solid-state regulator is a device called a *zener diode*. It looks and functions much like the standard diodes discussed in Chapter 2, but with one important difference. As we saw, a diode normally conducts current in one direction and blocks it (with the exception of a very slight leakage current) in the other direction. A zener diode works the same way in the forward or conducting direction and, up to a point, in the reverse or nonconducting direction. However, when this particular point is reached in the reverse direction, the zener diode experiences a nondestructive "breakdown." The diode no longer blocks

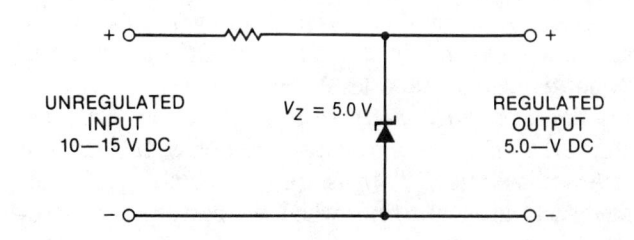

Fig. 3-16. Simple, zener-controlled voltage regulator.

29

Introduction to Automotive Solid-State Electronics

limiting resistor. As a result, the output voltage will be held at the zener voltage.

Such a circuit finds wide use in many electronic applications and forms an essential part of an automotive voltage regulator. Obviously, the simple circuit just described could not be used directly to regulate a charging system. But Fig. 3-17 shows how it might be incorporated into a regulator that can control the voltage in an actual system. Although circuit details differ among the various makes, practically all solid-state voltage regulators operate in a manner equivalent to the one we are about to discuss.

Voltage-Regulator Construction—A basic solid-state voltage regulator is shown in Fig. 3-17. If you have forgotten how transistors function, you may want to review that section in Chapter 2. The circuit shown consists of four major parts: (1) the voltage divider network of R_1, R_2, and P; (2) the zener diode D; (3) the driver transistor Q_1; (4) the output transistor Q_2. The purpose of the voltage divider is to scale down the system voltage to approximately the level of the zener voltage (about 5 V). The purpose of the driver transister Q_1 is to turn on and off the output transistor Q_2. Note that the alternator's field coil is connected into the collector–emitter circuit of the output transistor.

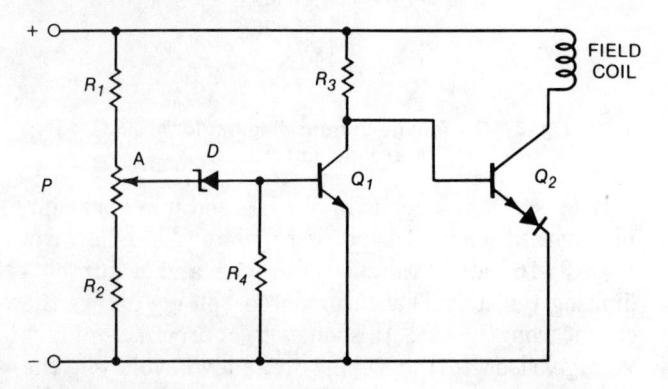

Fig. 3-17. Typical solid-state voltage-regulator diagram.

The potentiometer P in the voltage divider allows us to vary the regulated voltage over a certain range. This is helpful if we wish to "tailor" the voltage setting to a particular application or to alter it to meet different operating conditions. Many regulators do not have such an adjustable feature and are preset at some nominal voltage level.

Voltage-Regulator Operation—To start, let's assume that the system voltage is below the desired operating level. This is generally the case immediately after starting, when the battery is absorbing, for a few minutes, the full output of the alternator. Under these conditions, the voltage at point A is below the zener voltage of the diode. Since the diode is nonconducting, no current flows into the base of the driver transistor Q_1 and it remains cut off. This allows current to flow through R_3 and into the base of the driver transistor Q_2, thereby turning it on (the circuit is so designed that Q_2 is conducting heavily, i.e., it is in saturation). Full field current now flows, causing full output from the alternator.

As the battery comes up on charge, system voltage begins to rise. When it rises high enough (to the desired level), the voltage at point A exceeds the zener breakdown voltage of the diode and it conducts. The zener current flows into the base of the driver transistor and turns it on. The current that has been flowing in resistor R_3 (the base current for the output transistor) is now diverted through transistor Q_1. Without base current, the output transistor cuts off and field current stops. Alternator output drops and, consequently, system voltage decreases.

This brings us back to the initial conditions and the whole process starts over again. Just as with the vibrating voltage regulator, we control alternator output by switching the field current on and off. How fast this switching action takes place depends on the speed of the alternator and the demand for current by the battery and the electrical system. When the speed is high and the electrical load light, the switching may be occurring at a rate of several hundred times a second. At low speeds and/or heavy loads, the switching rate may be zero. That is, the field current is never switched off. Such conditions exist when the battery is in a discharged state or the electrical load equals or exceeds the output capacity of the alternator.

An actual regulator may have a few more components to enhance circuit operation and the arrangement of the basic elements may be somewhat different, but the overall functioning is practically as described. One component we have not shown, and which may be present in some regulators, is a temperature-compensation device to raise the voltage setting during cold operation. This is usually in the form of a *thermistor* inserted into the voltage-divider network. A thermistor is a temperature-sensitive resistor, i.e., one whose resistance value varies with temperature. By changing the ratio of the divider (through a temperature change), the regulated voltage setting is also changed.

CHAPTER 4

Ignition Systems

Of the various systems in the modern automotive engine, the ignition system seems to be the one surrounded with the most confusion. What many people think they know about ignition and what they actually know are frequently two different things. A lot of folklore has built up around the ignition system and has generated a lot of incorrect ideas concerning ignition theory. Let's see if we can sort out the facts from the fiction.

IGNITION CIRCUITRY

To start with, we consider the ignition process itself and later the problems of ignition timing. In actual practice, the two go hand in hand. Any ignition system, whether it be breaker-point, electronic, or capacitor discharge (the actual differences will be covered later), can be divided into two major circuits: (1) the primary circuit and (2) the secondary circuit.

Primary Circuit

The primary circuit is essentially a high-current, low-voltage circuit. Its main elements comprise the battery, the primary winding of the ignition coil, and a switching device. As we will see in a moment, the ignition coil is part of both the primary and secondary circuits. It is the term *switching device* that will require a little more explanation. We use this term to show that, in reality, there is little difference between so-called "electronic" ignition and the older "conventional" or breaker-point ignition. After we explain how a coil works, you will see that the difference is slight indeed.

Secondary Circuit

The secondary circuit is a high-voltage, low-current circuit. Its main elements include the secondary winding of the coil, the rotor, the distributor cap, the spark plug cables, and the spark plugs themselves. See Fig. 4-1. When we speak of high voltages, we mean voltages on the order of 5000 volts or more and currents in the milliampere range (1 milliampere = 0.001 ampere). When we speak of low voltages, we mean voltages of approximately 10-15 volts and currents of 1-4 amperes.

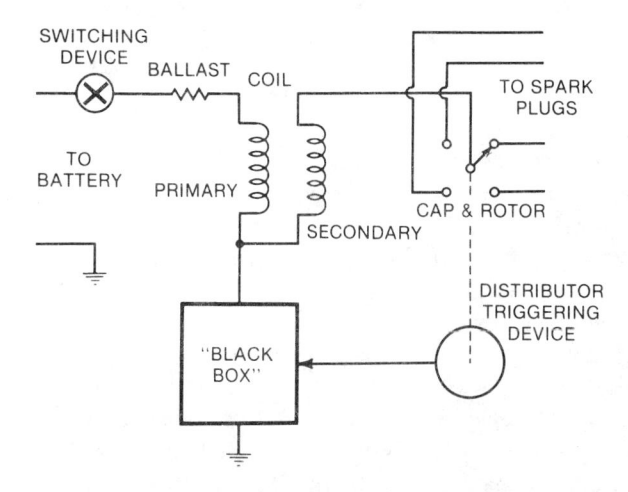

Fig. 4-1. Diagram of a "black box" ignition system. The black box can represent either mechanical breaker-points or a solid-state electronic switch.

Thus, we see that the ignition coil, being a part of both the low-voltage and the high-voltage circuits, becomes the "interface" between these two circuits. Its function is to transfer low-voltage, high-current energy into high-voltage, low-current energy. To see why this transformation is necessary, we must look at how a spark plug operates, since firing the plug is the end purpose of the ignition system.

SPARK PLUG OPERATION

There is a lot more to a spark plug than first meets the eye. But for now, consider it as nothing more than a spark gap—two electrodes separated by a certain distance with no apparent connection between the two electrodes. In low-voltage circuits, the two electrodes act as an open circuit; i.e., no current would flow between them. In a high-voltage circuit, the situation changes. We have to consider an effect known as *ionization*.

Air is normally considered to be a nonconductor of

Introduction to Automotive Solid-State Electronics

electricity. However, as lightning so clearly demonstrates, when enough voltage is impressed across the air, it breaks down and conducts. The visual effect of this conduction is a spark or, during the course of an electrical storm, a lightning bolt. What we are witnessing, of course, is the result of the process of ionization. Without ionization there would be no conduction or spark discharge.

Almost any gaseous substance can be ionized, including the air-fuel mixture in the cylinder of an engine. All it takes is one free *ion*. An ion is a positively or negatively charged atom or molecule that exists naturally in the air and other gaseous substances. Additional ions can also be created when ionized particles collide with nonionized particles. When enough ionized particles exist between two electrodes, they serve as a "current bridge" to carry current from one electrode to the other—a miniature lightning bolt.

We can ionize a spark gap, and thus prepare it to conduct current, by applying a sufficiently high potential difference to the electrodes of the gap. When a sufficiently high voltage, called the *ionization potential*, is impressed across a spark gap, any free ion present in the gap will be attracted to one or the other electrode. As the ion moves, it collides with other particles, ionizing them in the process. These in turn are accelerated toward the electrodes and collide with other particles, forming even more new ions. The process is cumulative and soon the gap is sufficiently ionized to "break down" or conduct current. The visual result is a spark. Technically, we have simply dissipated a quantity of energy across the gap. If a combustible mixture of air and fuel is present, we will have initiated the combustion process. Note that if a free ion had not been present to start with, no ionization (and hence no ignition) would have taken place. Without ignition, there is no combustion. This would have been a misfire.

To summarize, the combustion process cannot begin until there is ignition—i.e., the dissipation of electrical energy across the spark gap. Electrical energy (manifested by the flow of current across the gap) cannot occur unless the gap is ionized. The gap cannot be ionized unless there is (1) a free ion to start with and (2) a sufficiently high voltage across the gap to accelerate the ion enough to produce more ions. The whole process is a chain reaction; if any link is missing, the necessary reactions cannot happen and ignition does not occur. As we said before, this results in a misfire. Before we go on, let's consider some of the pitfalls that can cause problems in the ignition process.

IGNITION PROBLEMS

What is the probability that a free ion will not be available at the moment we wish to have ignition? There is a slight but nevertheless very real possibility that this can happen, especially when an engine is operating under very lean air-fuel mixtures (i.e., considerable air and little fuel). You are more likely to find a free ion in fuel than in air. Modern engines are operating much leaner than they did a few years ago, so the probability of a lack of free ions has increased. To counteract this, car makers have widened the spark gap. The logic behind this practice is obvious: The wider the spark gap, the greater the probability that an ion will be present between the gap. The practical limits to the width of the gap will be discussed later.

What is the probability that a combustible mixture will not be present at the gap even though a spark occurs? Again, this becomes more of a possibility as mixtures become leaner. There is only a certain range of air-fuel ratios (AFR) over which reliable combustion can occur, typically from about 8:1 AFR (very rich) to about 18:1 AFR (very lean). Mixtures outside this range are either too rich or too lean to support combustion. One technique to minimize misfires due to leanness is to use wider spark plug gaps (which also helps in the ionization process, as mentioned above). The wider the gap, the greater the possibility of a combustible mixture being present within the gap. (Note: The air-fuel mixture is not a homogeneous mixture evenly distributed throughout the cylinder. Depending on compression turbulence and other factors, there can be pockets of very lean, or very rich, mixtures scattered about, even though the overall mixture ratio is normal.)

What is the probability that there will be insufficient voltage to ionize the plug gap? This is the most common reason for ignition misfiring since it can be caused by a number of conditions. Unless we can impress enough voltage across the gap, we cannot initiate the ionization process. And without this, there can be no spark or ignition. But voltage alone is not the whole answer; how voltage is applied, or rather concentrated, is what really counts. To better illustrate this, let's make an analogy.

Suppose you wished to walk across a frozen pond. If you were to cross it on snowshoes, you would probably have no problem. But try to do it on stilts and you would most likely break through the ice. The reason is obvious: Stilts concentrate all of your weight on a small area, whereas snowshoes distribute it evenly over a wide area.

Ignition Systems

We can use the same reasoning in considering the voltage applied to a spark plug gap. Only in this case we wish to concentrate the voltage as much as possible, since the object is to "break through" or ionize the normally nonconductive gap. Voltage across a gap can be concentrated or "focused" by sharp edges or points. This is illustrated in Fig. 4-2. When the electrodes of the spark gap become rounded by wear and erosion, they lose their ability to concentrate the impressed voltage. This means that the applied voltage must be much higher than normal in order to achieve the necessary ionization potential. Since the maximum available voltage from the ignition system is not unlimited, it may never reach the required ionization potential. The result will be a misfire.

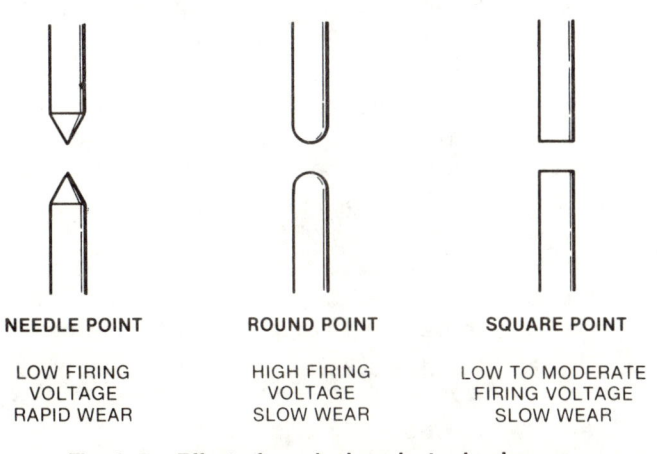

NEEDLE POINT — LOW FIRING VOLTAGE, RAPID WEAR

ROUND POINT — HIGH FIRING VOLTAGE, SLOW WEAR

SQUARE POINT — LOW TO MODERATE FIRING VOLTAGE, SLOW WEAR

Fig. 4-2. Effect of spark-plug electrode shape on sparking or firing voltage.

There are other reasons, as we will see later, that can cause a lack of high voltage (ionization potential). But for now let's see how the ignition system develops this high voltage. Fig. 4-1 illustrated the major elements of a basic ignition system. Note that we do not refer to this system as either electronic or breaker-point (sometimes called conventional): it could be either. Instead, we identify the controlling element as a *switching device*. If you look upon ignition systems in this manner, there is very little difference between modern electronic ignition and earlier systems.

IGNITION COIL

We mentioned previously that the coil was the interface between the primary circuit (the low-voltage portion) and the secondary circuit (the high-voltage portion). The ignition coil is basically a transformer: it transforms low-voltage energy into high-voltage energy. Just how it does this will become clear if you recall the principles by which an alternator or generator operates. Remember, all it takes to generate a voltage is (1) a conductor, (2) a magnetic field, and (3) motion.

All three of these elements are present in an ignition coil. The coil contains two windings (our conductors)—a primary winding and a secondary winding. The primary winding is essentially an electromagnet and if we pass current through it, we create a magnetic field—the second element needed to generate voltage. But how do we get motion, the necessary third element?

Simple. We just collapse the magnetic field by abruptly stopping the primary current that created it in the first place. This is the job of the switching device, whatever it may be. As the magnetic field collapses, its magnetic lines of force cut across the conductors of the secondary winding and a voltage is induced in each turn. Since there are so many turns in the secondary (approximately 20,000 compared to about 200 in the primary), a very high voltage will be generated. This is how we get the necessary ionization potential to "fire" the plug gap. See Fig. 4-3.

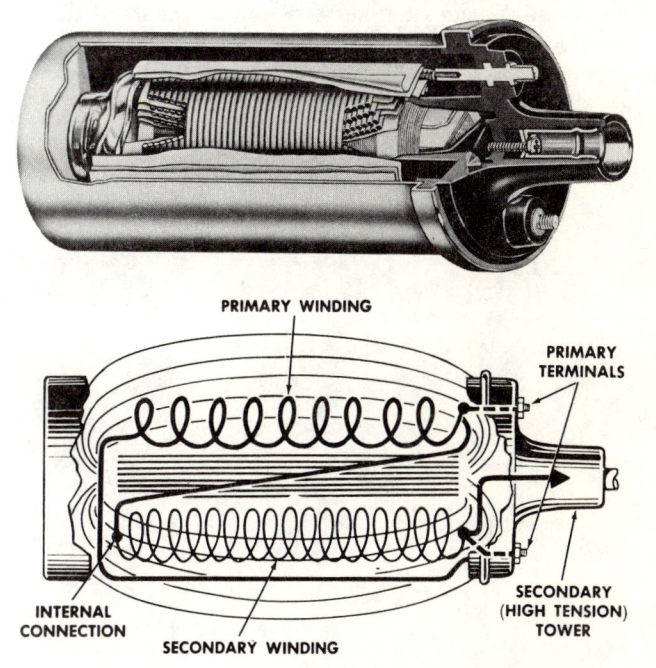

PRIMARY WINDING
PRIMARY TERMINALS
INTERNAL CONNECTION
SECONDARY WINDING
SECONDARY (HIGH TENSION) TOWER

Courtesy C. E. Niehoff and Company

Fig. 4-3. Magnetic field within the ignition coil.

The whole secret is to collapse the magnetic field as rapidly as possible. If the field collapses slowly, its effective motion will be slow and, consequently, the induced voltage will be low. The same thing will occur if the field does not collapse completely, a distinct possibility with solid-state ignition systems.

Introduction to Automotive Solid-State Electronics

Condenser

In the earlier breaker-point or so-called conventional ignition systems, the switching device was simply a set of electrical contacts made to open and close by the distributor cam. When they closed, current flowed through the primary winding, creating the magnetic field. When they opened, the current stopped and the magnetic field collapsed. This generated the high voltage we have just described. The problem with breaker-point ignition, was that the primary current did not want to stop flowing after the contacts opened. If not checked, it would tend to keep on flowing by creating an arc across the contacts. This was highly undesirable since it not only reduced the voltage output of the coil, but also burned the contacts. To remedy this situation, a condenser was connected across the contact points. The condenser acted as a surge chamber to temporarily check the current that would have been arcing over the contacts as they separated. The net effect was to bring the primary current to a rapid and controlled halt, thus assuring a high voltage output from the coil. The condenser did not eliminate all contact arcing (a slight amount had a beneficial cleaning action on the contacts) and in time the contacts wore out through electrical erosion. See Fig. 4-4.

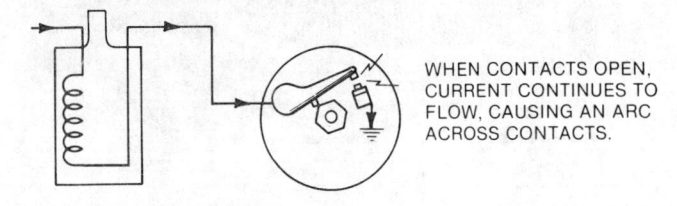

(A) No condenser in primary circuit.

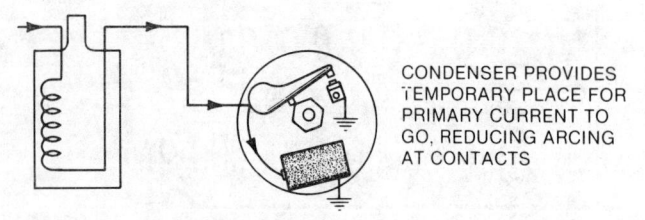

WHEN CONTACTS OPEN, CURRENT CONTINUES TO FLOW, CAUSING AN ARC ACROSS CONTACTS.

(B) Condenser in primary circuit.

CONDENSER PROVIDES TEMPORARY PLACE FOR PRIMARY CURRENT TO GO, REDUCING ARCING AT CONTACTS

Fig. 4-4. Action of the condenser in a breaker-point system

Contrary to popular belief, condensers never wore out. Sometimes they would fail because the insulating material between their plates would break down or the internal connections would corrode and impede condenser action. But such failures were quite rare. The vast majority of condensers were replaced out of habit rather than need.

Solid-State Switches

Solid-state switches have taken the place of the mechanical breaker-points in present domestically produced vehicles. Although a solid-state switch (transistor) operates differently from mechanical contacts, the results are exactly the same. As we have seen in earlier chapters, a transistor can turn current on and off by means of a small control current injected into its base. By controlling this base current, we can switch primary current on and off as we choose. When primary current is switched off, the coil's magnetic field collapses and a high voltage is generated in the secondary circuit. Does this sound familiar? It should: it is exactly the same action that took place with breaker-point ignition. This is why we can call the device that turns the primary current on and off a switching device. We really don't care what it is as long as it can switch primary current.

However, solid-state switching offers certain advantages over the earlier breaker-point system. For one thing, we can eliminate the ignition condenser, since transistor switches do not have the problem of arc-over as do mechanical contacts. This is a rather minor point since it takes 4-6 condensers elsewhere in the circuit to make the transistor switch work properly. A transistor can handle much more current than mechanical breaker-points. To see why this is important, we must consider some additional aspects of ignition systems.

High-Voltage Energy Production and Limitations

A coil cannot produce more high-voltage energy than what was put into it. The amount of energy that can be put into a coil depends on (1) how much primary current is flowing at the moment it is interrupted and (2) the *inductance* of the primary winding. Increasing either or both of these increases the input (and consequently the output) energy. Inductance is easily increased simply by adding more turns to the primary winding. Primary current can be increased by reducing the resistance of the primary circuit. It would seem that we could get as much energy out of a coil as we desired.

There is a practical limit, however, to just how much energy a coil can produce. With breaker-point ignition we were limited by how much primary current the contacts could safely handle, typically about 3-5 amperes. Increasing the inductance by adding more turns of wire would increase the energy output, but at the sacrifice of coil saturation time. As shown in Fig. 4-5, coil primary current does not reach its maximum level until some time after it starts to flow.

34

Ignition Systems

This is the coil's saturation time which has a direct effect on the coil's output voltage. If the primary current is interrupted before the coil saturates, coil output will be reduced. And this is precisely what can happen under high-speed operation. Unfortunately, high-speed operation frequently requires more, not less, coil output.

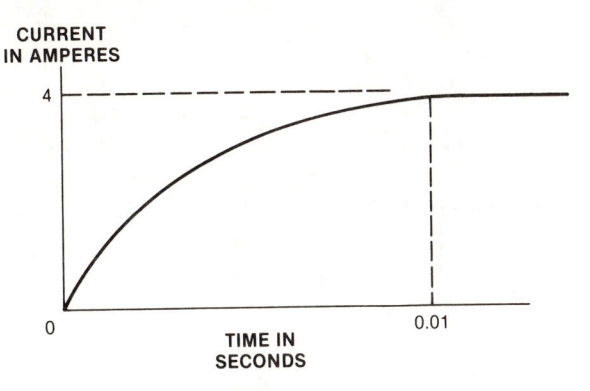

Fig. 4-5. Ignition primary current buildup in a coil's primary winding.

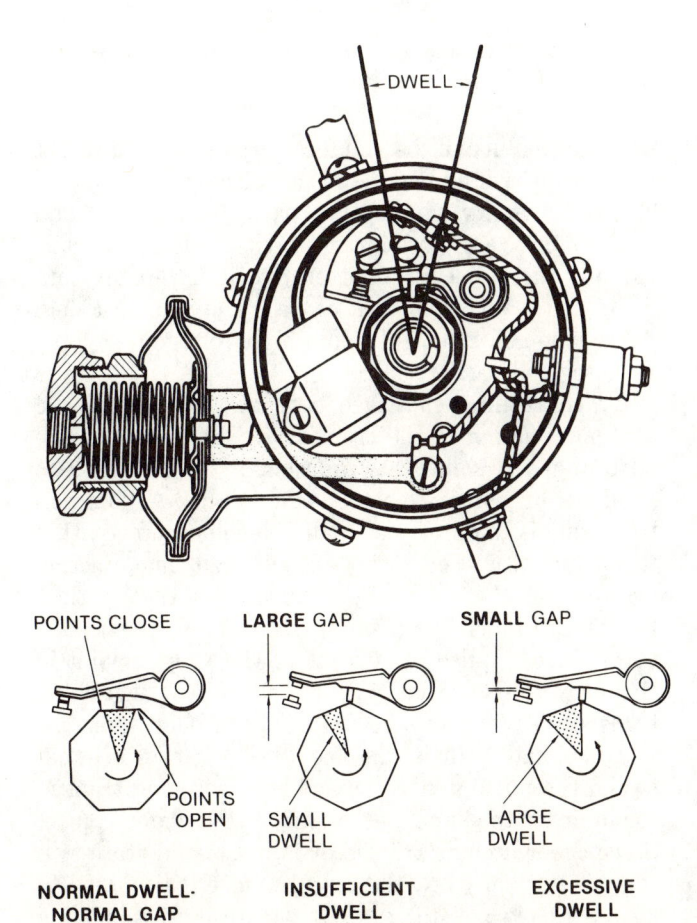

Fig. 4-6. Cam or dwell angle in an actual distributor.

Solving Coil Output Reduction

Breaker-point systems tackled this problem several ways. One was to use long dwell periods to achieve better coil saturation at high speeds, sometimes using dual overlapping points to further extend the dwell period. (Dwell is the angular distance of distributor cam rotation during which the points are closed and primary current is flowing. See Fig. 4-6.) Another technique was to use a ballast resistor in series with the primary winding. For reasons we won't go into here, placing a resistance in series with an inductance will decrease the saturation time of the inductance. The ballast resistor, which is still employed in certain solid-state systems, also provided an additional advantage. It could be bypassed or shorted out during cranking to compensate for reduced battery voltage, thus maintaining adequate coil output for starting.

Solid-state systems are not as limited by these constraints. A transistor can switch a much higher current than a set of mechanical breaker-points. Because of this, the coil's primary inductance can be reduced for better saturation at higher speeds. The higher switching current of the transistor makes up for the reduced inductance of the coil. As a result, coil output with a solid-state system can be even higher than that of a breaker-point system. This is essentially what is done in so-called High Energy Ignition systems.

IGNITION DEMAND

Now that we have considered the role of the ignition coil in producing high-voltage energy, the concept of *ignition demand* is introduced. Just because a coil can produce a maximum of 30,000 volts (called "available voltage") does not mean that it will always deliver that much to the spark plugs. The coil will produce only as much voltage as the spark plugs demand. This fact is contrary to the popular notion that a coil always delivers its maximum voltage to the plugs. The hottest coil you can buy will develop no more voltage than it takes to fire the plug. This then raises the question, "How much voltage does it take to fire a plug?"

The answer is subject to many qualifications, since many factors affect the firing voltage of a spark plug. The following factors are the major ones.

(1) Plug gap—in general, the wider the gap, the higher the firing voltage.
(2) Condition of the plug electrodes—this has considerable effect on firing voltage. Worn or rounded electrodes require much higher firing voltages.

Introduction to Automotive Solid-State Electronics

(3) Engine compression ratio—as the compression ratio goes up, so does the firing voltage. Also, as the compression pressure increases (from part throttle to wide-open throttle), so do the firing voltages.

(4) Air–fuel ratio—lean mixtures require higher firing voltages than rich mixtures.

As you can see, there is no simple answer to how much voltage it takes to fire a plug. It can vary from less than 10,000 volts to over 30,000 volts in a typical engine under normal driving conditions. The point to remember is that the coil produces only as much voltage as the plug demands, assuming, of course, that the plug does not demand too much. When the plug demands more voltage than the coil can produce, a misfire occurs. We literally "run out of ignition."

The difference between the maximum available voltage from the coil and the required voltage of the plugs is called *ignition reserve*. As long as the required voltage is less than the available voltage there is no problem. Ignition systems are designed to have an adequate ignition reserve to allow for unusual conditions and normal plug wear and deterioration. But when excessive wear takes place, normal ignition reserve cannot compensate and missing occurs. Missing frequently shows up first during part-throttle acceleration, since this mode usually imposes the greatest demands on the plugs.

SWITCHING DEVICES

So far, we have studied the process of ignition and the many factors involved in simply firing a spark plug. As you can see, it is a many-faceted problem and perhaps not as simple as it first appeared. This brings us to the final topic in our discussion: the switching device. We stated earlier that the switching device could be either a set of mechanical contacts (breaker-points) or a solid-state switch (a transistor). It makes little difference to the coil and the rest of the ignition system which device is used. To operate the coil requires only that the primary current be turned on and off at the proper times.

Since we have already covered the operation of breaker-points (primarily from an historical standpoint), we can direct our attention to the solid-state switch. The main element in this circuit is a power transistor that can handle the necessary primary current. A basic solid-state circuit is shown in Fig. 4–7. In Chapter 2 we showed how a transistor could be made to function as a switch. When a sufficient current is caused to flow in the transistor's base circuit, the collector–emitter circuit will be driven into saturation; i.e., it will be conducting as though a direct connection existed between the collector and emitter. The coil primary winding is a part of this circuit and is therefore receiving current. When base current is cut off, the collector–emitter circuit no longer conducts and primary current ceases. The magnetic field within the coil collapses and a high voltage develops, as described earlier.

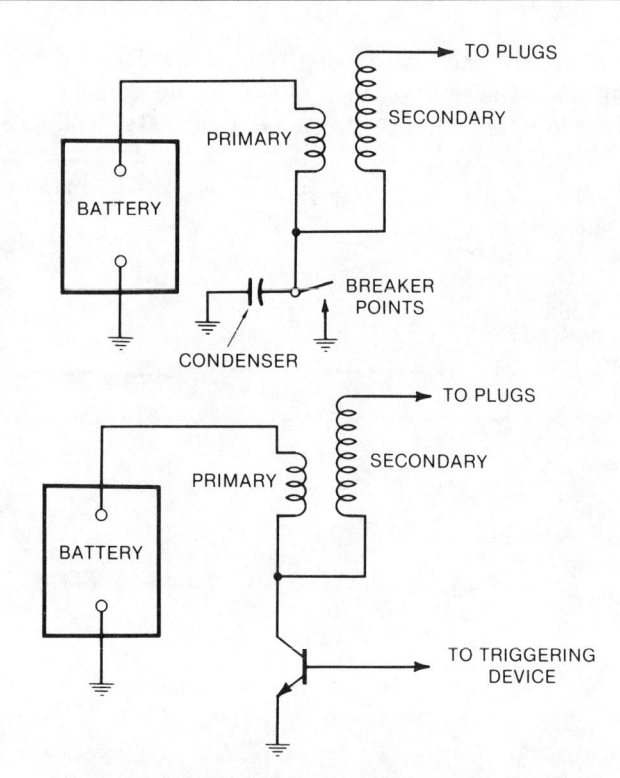

Fig. 4–7. Comparison of basic solid-state ignition to breaker-point system.

By controlling the base current, we control the switching action of the transistor. This is one of the advantages of solid-state ignition. It is quite easy to produce or control this base current. A common technique is to employ a small pulse generator within the distributor. See Fig. 4–8. Different manufacturers refer to these by different names, although they function in much the same manner. Their purpose is to produce a pulse of current at the proper point in the engine cycle that will switch off the switching transistor. When this occurs, a spark plug fires.

Additional transistors are used to interface between the pulse generator and the switching transistor in order to shape and amplify the "trigger" pulse. Since the switching transistor in most solid-state systems is normally ON, the trigger pulse is designed to switch this transistor OFF. It does this by removing the base current from the transistor. This action is represented in Fig. 4–9. A four-cylinder reluctor is

Ignition Systems

shown. Transistor Q_1 amplifies the feeble trigger pulse from the pickup coil and drives the switching transistor Q_2.

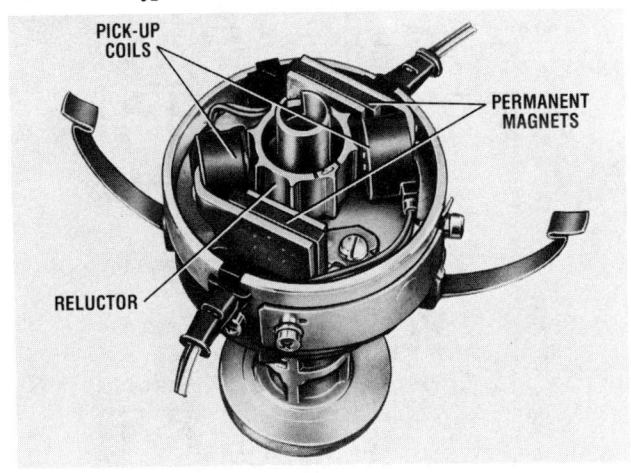

Fig. 4-8. Distributor with the reluctor or magnetic-pulse generator type of pickup system.

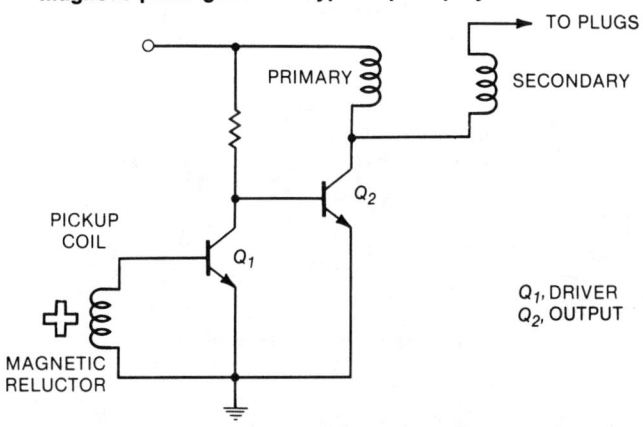

Courtesy Chrysler Corporation

Fig. 4-9. Reluctor pickup system coupled to a basic solid-state ignition system.

The timing of an ignition system using the pulse generator can be accomplished in the same manner as a breaker-point system; that is, by the use of vacuum and mechanical advance mechanisms. See Fig. 4-10. However, it is also possible to generate the necessary trigger pulse by means of a computer. This allows the ignition timing function to be accomplished by purely electronic devices rather than by mechanical devices. Since the amount of timing advance needed at any particular instant can be evaluated through a variety of computer inputs, much greater timing precision can be obtained than would otherwise be possible. For example, one of the computer inputs might be a knock detector. This, coupled with other inputs, can help maintain the spark timing close to the threshold of knock, thereby holding timing close to the op-

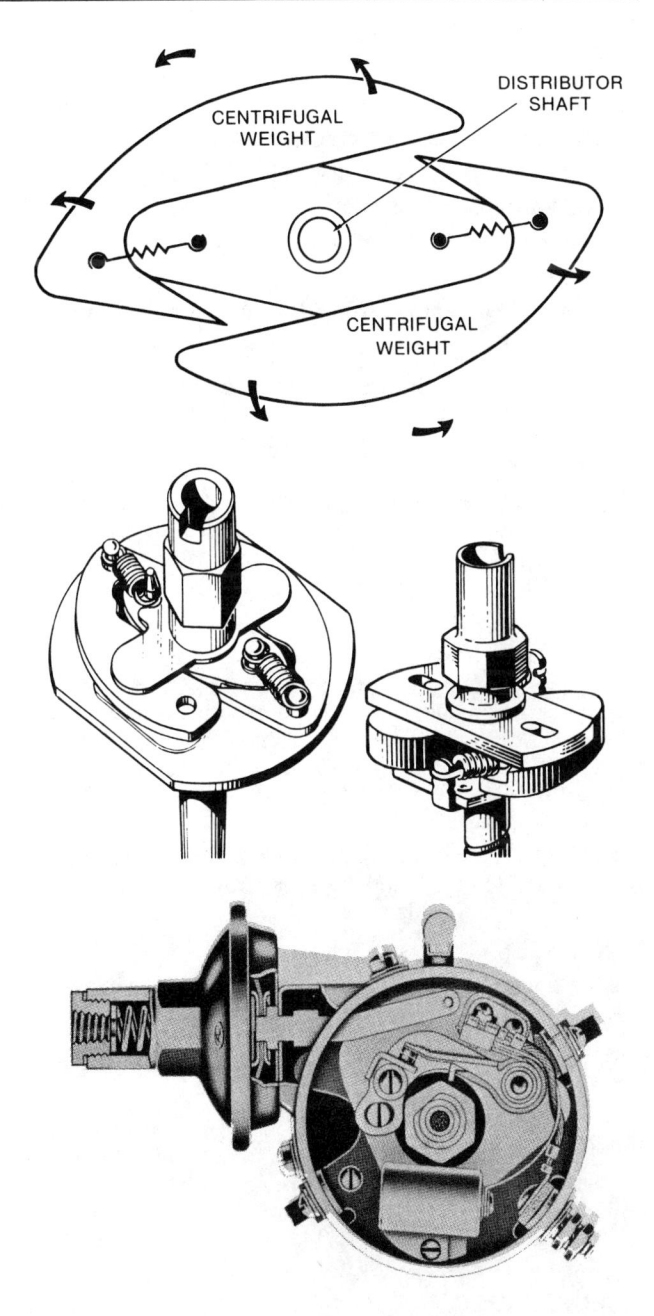

Fig. 4-10. Vacuum and mechanical advance mechanisms.

timum operating point. (See Chapter 8 for greater detail on Knock Limiters.)

TROUBLESHOOTING CHARTS AND SERVICE DATA FOR ELECTRONIC IGNITION SYSTEMS

Figs. 4-11, 4-12, 4-13, and 4-14 give you an idea of troubleshooting procedures for American Motors, Chrysler, General Motors, and Ford electronic ignition systems.

Introduction to Automotive Solid-State Electronics

BEFORE YOU BEGIN:

- Check Fuel Delivery.
- Check Battery Cranking Voltage — 9.5 Volts Or More.
- Check All Primary and Secondary Wiring For Damage and Loose Connections.
- Ignition On, Crank Engine, Check For Spark at Spark Plug.
- Use 12-Volt Test Lamp (No. 57 Bulb) and Ohmmeter For These Tests.

1. Remove coil wire from distributor cap and hold it ½ inch from ground. Crank the engine and watch for steady spark. If spark occurs, ignition primary is OK. The problem is in the cap, rotor or cables.

2. If no spark occurs, connect test lamp from positive (+) terminal on coil to ground. Turn ignition switch to both ON and START. The lamp should light in both positions. If not, fault is in circuit between battery and coil.

CONTROL UNIT TEST

3. If lamp lights in both positions of Step 2, connect lamp across positive (+) and negative (−) terminals of coil. Unplug the sensor leads near distributor and turn ignition ON. If lamp does not light, check control unit ground connection. If ground is OK, control unit is defective and must be replaced.

4. If lamp lights in Step 3, short across terminals in sensor lead coming from control unit. If lamp does not go off, fault is in control unit.

COIL TEST

5. If lamp goes off in Step 4, remove lamp and re-establish ½ inch gap from coil wire to ground. Short across terminals of sensor lead as in Step 4. If spark does not occur each time terminals are shorted, coil is faulty.

6. With ignition OFF, use an ohmmeter to measure coil primary resistance between positive (+) and negative (−) terminals. The meter should read 1.25 to 1.40 ohms. Measure secondary resistance between center tower and positive (+) coil terminal. The meter should read 9,000 to 15,000 ohms. If either test is not within specifications, replace coil.

SENSOR TEST

7. If spark occurs from coil when sensor terminals are shorted in Step 5, fault is in sensor or control unit. Using an ohmmeter, measure sensor resistance at distributor leads. The meter should read 1.6 to 2.4 ohms at 77° to 200° F. Tap sensor lightly; resistance should not vary. If resistance is not within limits, replace sensor.

8. Connect one ohmmeter lead to either sensor lead; touch other lead to center core of sensor. The ohmmeter must show open circuit.

9. If sensor passes these tests, the problem is most likely in the control unit. Replace control unit and retest.

Courtesy Champion Spark Plug Company

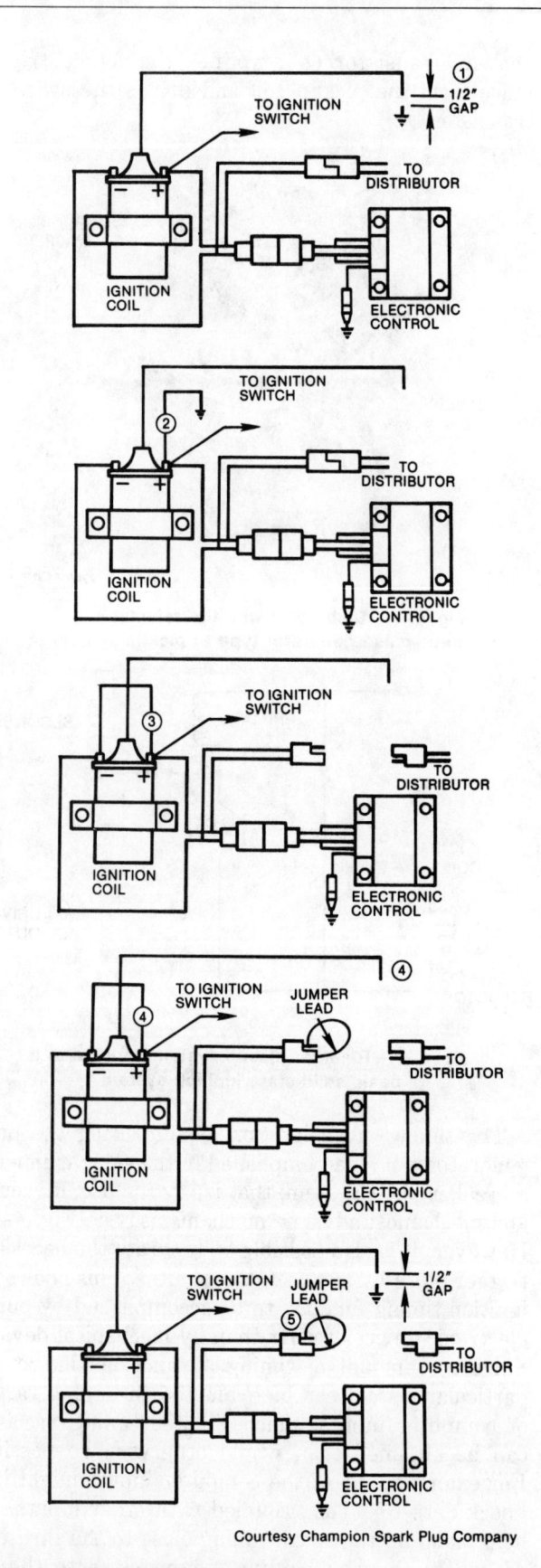

Fig. 4–11. American Motors breakerless inductive discharge ignition system.

Ignition Systems

BEFORE YOU BEGIN:

- Check Fuel Delivery.
- Check Battery Cranking Voltage — 9.5 Volts Or More.
- Check All Primary and Secondary Wiring For Damage Or Loose Connections.
- Ignition On, Crank Engine, Check For Spark at Spark Plug.
- Check Air Gap Between Reluctor and Pickup Coil With Non-Magnetic Feeler Gauge. Gap should be .008".

CAUTION: *Do not touch transistor on control unit when ignition is ON. Sufficient voltage is present to produce severe shock.*

CAUTION: *Ignition switch must be OFF when connector is removed from, or connected to, the control unit.*

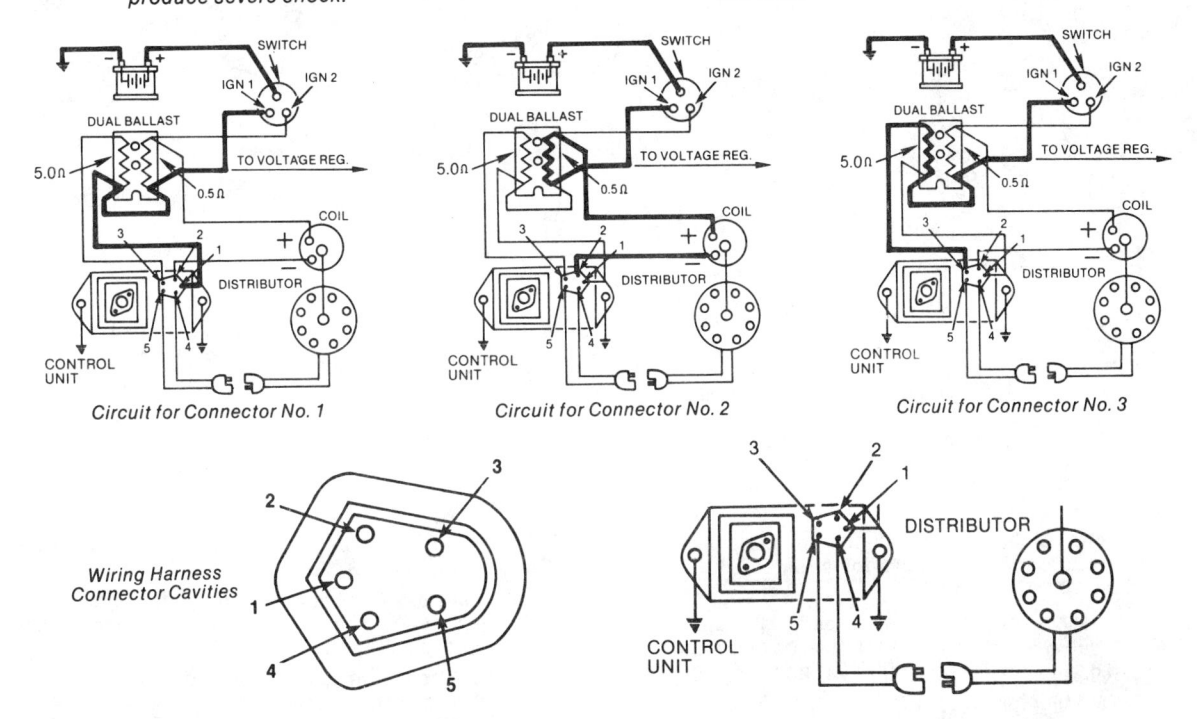

Circuit for Connector No. 1 *Circuit for Connector No. 2* *Circuit for Connector No. 3*

Wiring Harness Connector Cavities

1. With ignition OFF, remove wiring harness connector from ignition control unit.
2. Turn ignition ON and connect voltmeter negative lead (−) to ground.
3. Connect voltmeter positive (+) lead to harness connector cavity #1. With ignition ON and all accessories OFF, voltage should be within 1 volt of battery voltage. If not, trace and repair circuit illustrated.
4. Repeat Step 3 at harness connector cavity #2. If voltage is not within 1 volt of battery voltage, trace and repair circuit illustrated.
5. Repeat Step 3 at harness cavity #3. If voltage is not within 1 volt of battery voltage, trace and repair circuit illustrated.

PICKUP COIL TEST

6. With ignition OFF, connect ohmmeter to harness connector cavities #4 and #5. Pickup coil resistance should read between 150 and 900 ohms.
7. If the pickup coil resistance is out of limits in Step 6, disconnect dual-lead connector from distributor and check resistance through pickup coil leads at distributor.

8. If reading in Step 7 is still out of limits, replace pickup coil. If reading is within limits, check wiring from control unit to distributor.
9. Connect one ohmmeter lead to ground and the other lead alternately to both sides of the pickup coil, dual-lead connector, at distributor. The ohmmeter should show open circuit. Any continuity indicates grounded pickup coil.

GROUND CIRCUIT TEST

10. Connect one ohmmeter lead to ground and the other to control unit connector pin #5. The ohmmeter should show continuity between ground and pin #5. If not, tighten control unit mounting bolts and retest. If continuity still does not exist, replace control unit.

COIL AND BALLAST RESISTOR RESISTANCE TESTS

11. With ignition OFF, use the ohmmeter to measure coil primary resistance between positive (+) and negative (−) terminals. The meter should read 1.41 to 1.79 ohms at 70-80° F.
12. Use the ohmmeter to measure resistance across both sides of dual ballast resistor. The normal side should read .5 to .6 ohms. The auxiliary side should read 4.75 to 5.75 ohms. If either side is not within specifications, replace the unit.

Courtesy Champion Spark Plug Company

Fig. 4-12. Chrysler electronic ignition system.

Introduction to Automotive Solid-State Electronics

BEFORE YOU BEGIN:
- Check Fuel Delivery.
- Check Battery Cranking Voltage — 9.0 volts or more.
- Check All Primary and Secondary Wiring For Damage and Loose Connections.
- Ignition On, Crank Engine, Check For Spark at Spark Plug.
- Use Voltmeter and Ohmmeter For These Tests.

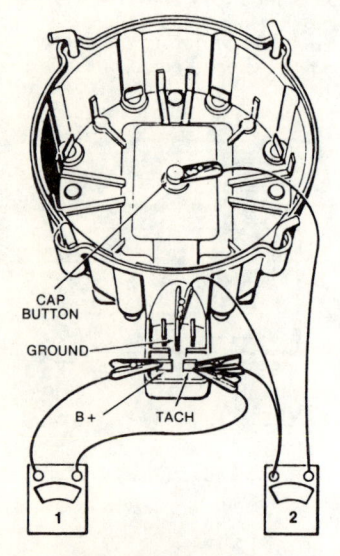

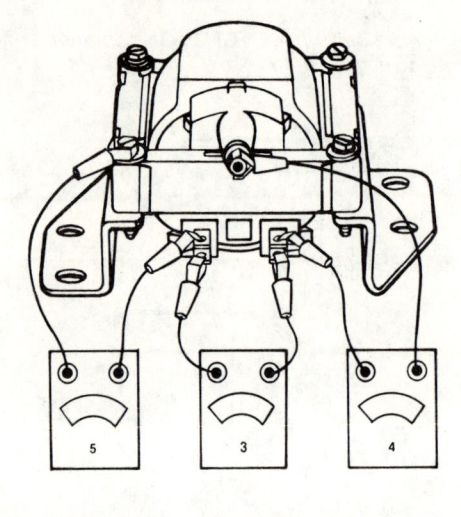

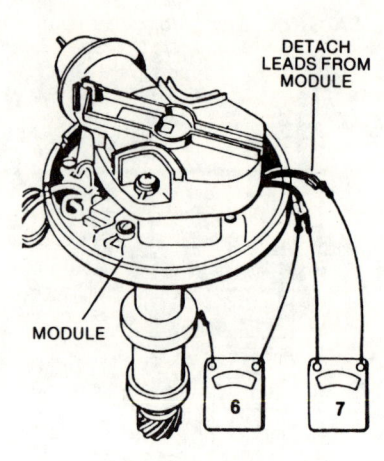

HEI V-8 and V-6 Coil Test Points

HEI 4- and 6-Cylinder Coil Test Points

HEI System Pickup Coil Test Points

1. Be sure that all four distributor cap latches are secure and that coil and coil cover screws are tight on V-6 and V-8 units.
2. Connect voltmeter negative lead (−) to ground and positive lead (+) to BAT terminal on distributor. (4, 6 & 8 cyl. w/coil in distributor cap only).
3. Turn ignition switch ON.
4. The voltmeter should read approximate battery voltage. If reading is zero, circuit is open between battery and distributor. Trace and repair as required.

IGNITION COIL TEST

5. Remove cap and coil assembly from V-6 and V-8 distributors by releasing four latches. Coils for in-line 4 and 6 cylinder systems are mounted separately from distributor on earlier production cars.
6. Visually inspect cap, coil and rotor for signs of cross firing and leakage.
7. Make resistance tests with an ohmmeter, as follows, for each system:

V-6 AND V-8 HEI SYSTEMS; 1978 AND LATER IN-LINE 4 AND 6 CYLINDER SYSTEMS (WITH COIL IN DISTRIBUTOR CAP):

a) Test point 1 measures primary resistance. Ohmmeter should show continuity (0 to 1 ohm). If not, replace coil.

b) Connect ohmmeter both ways, Step 2. Use high scale. Replace coil *only if both readings are infinite.*

1975-'77 IN-LINE 4 AND 6 CYLINDER SYSTEMS (WITH COIL MOUNTED SEPARATELY):

a) Test point 3 measures primary resistance. Ohmmeter should show continuity (0 to 1 ohm). If not, replace coil.

b) Test point 4 measures secondary resistance. If reading is more than 30,000 or less than 6,000 ohms, replace coil.

c) Test point 5 should show infinite resistance (open circuit). If resistance is less than infinite, replace coil.

HEI PICKUP COIL TESTS

8. Connect a vacuum source to vacuum advance unit and check vacuum advance operation. If vacuum unit is inoperative, replace it.
9. Detach two pickup coil leads from control module in distributor.
10. Connect ohmmeter to test point 6 and operate vacuum advance through full range.
11. Resistance between either of the pickup coil leads and the distributor housing must be infinite. If not, replace pickup coil.
12. Connect the ohmmeter to pickup coil leads (test point 7) and operate vacuum advance through full range. If ohmmeter reads less than 500 or more than 1500 ohms, replace pickup coil.
13. If no defects have been found in the ignition coil or pickup coil and ignition is still faulty, check the module with an approved module tester. Follow module tester manufacturer's directions. If module is faulty, replace.

Courtesy Champion Spark Plug Company

Fig. 4-13. Delco (GM) (HEI) system.

Ignition Systems

BEFORE YOU BEGIN:
- Check Fuel Delivery.
- Check Battery Cranking Voltage — 9.5 Volts Or More.
- Check All Primary and Secondary Wiring for Damage and Loose Connections.
- Ignition On, Crank Engine, Check For Spark at Spark Plug*.
- Use Voltmeter, Ohmmeter, and Jumper Wire For Tests.
- Follow Chart and Diagrams For Year Of Vehicle.

*CAUTION: For Spark Test, Do Not Pull These Cables While Engine Is Running:
No. 1 or No. 8 on V-8 Engines
No. 3 or No. 5 on In-Line 6 Engines
No. 1 or No. 4 on V-6 Engines
No. 1 or No. 3 on 4-Cylinder Engines

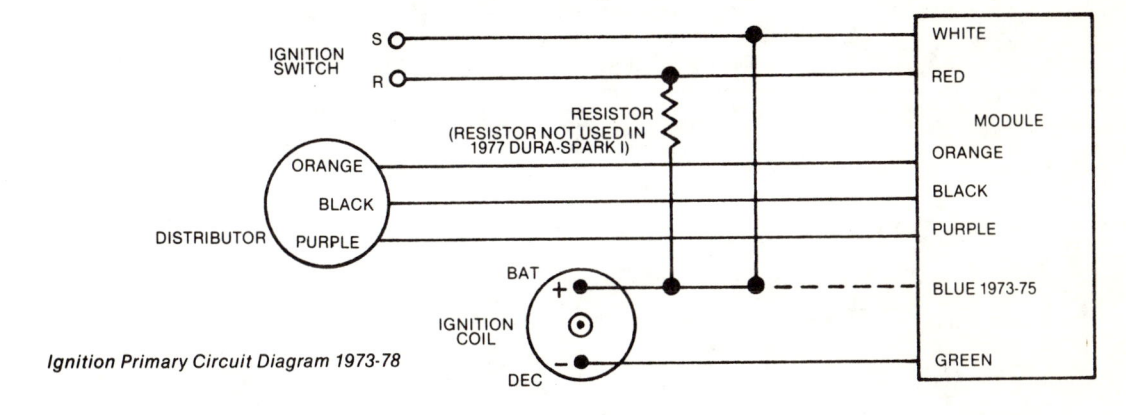

Ignition Primary Circuit Diagram 1973-78

1973-74 SOLID STATE

	Test Voltage Between	Should Be	If Not
Key On	Coil BAT terminal and ground. (Module connected, DEC terminal grounded)	4.9 to 7.9 volts	Low reading - Check primary wiring. High reading - Replace resistance wire
	Socket 3 (Red) and ground	Battery voltage ± 0.1 volt	Repair red wire, check connectors
	Socket 5 (Green) and ground	Battery voltage ± 0.1 volt	Check green wire to coil, check coil
Cranking	Socket 1 (White) and ground	8 to 12 volts	Repair white wire, check connectors
	Socket 5 (Green) and ground	8 to 12 volts	Check green wire to coil, check bypass circuit
	Socket 7 (Purple) and Socket 8 (Orange)	0.5 volt ac or any dc voltage	Replace magnetic pickup (stator)
	Test Resistance Between	**Should Be**	**If Not**
Key Off	Socket 7 (Purple) and Socket 8 (Orange)	400 to 800 ohms	Replace magnetic pickup (stator) or repair ground connection
	Socket 6 (Black) and ground	0 ohms	
	Socket 7 and ground	more than 70,000 ohms	
	Socket 8 and ground	more than 70,000 ohms	
	Socket 3 (Red) and coil tower	7,000 to 13,000 ohms	Replace coil
	Socket 5 (Green) and Socket 4 (Blue)	1.0 to 2.0 ohms	
	Socket 5 (Green) and ground	more than 4.0 ohms	Check for short at coil DEC terminal or in wiring to DEC terminal
	Socket 3 (Red) and Socket 4 (Blue)	1.0 to 2.0 ohms	Replace resistance wire

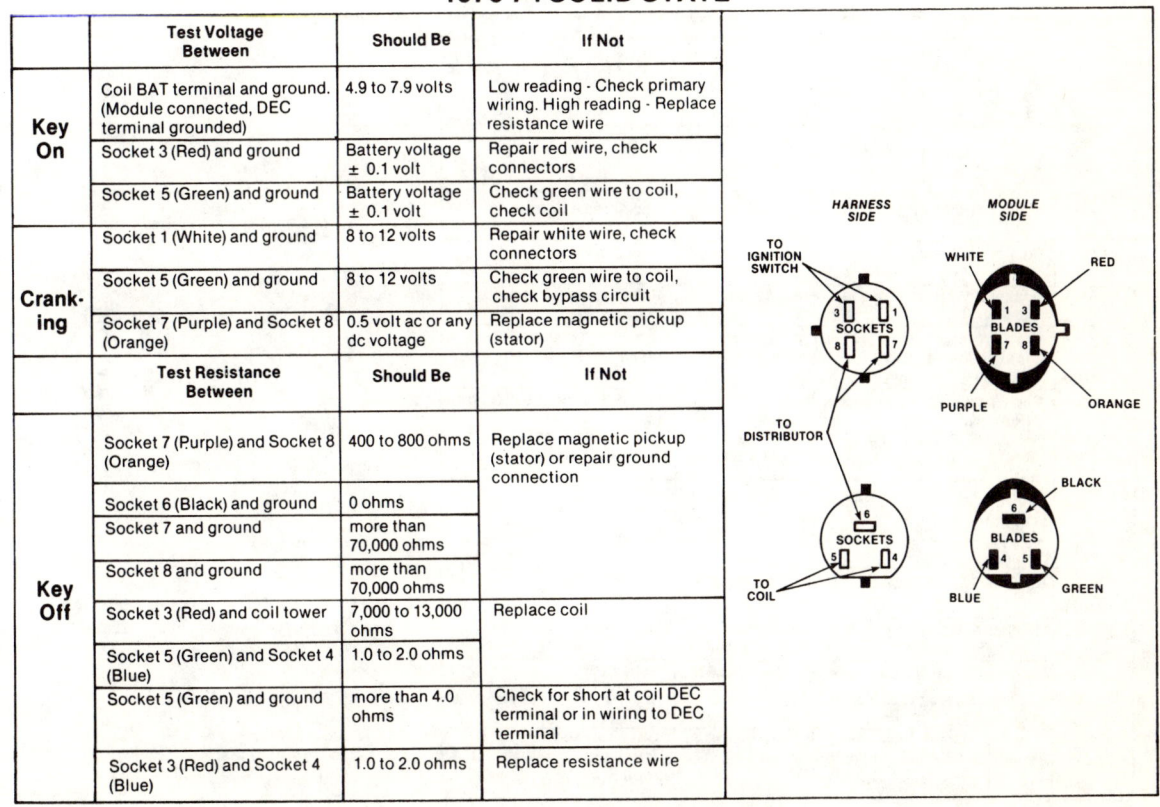

Fig. 4–14. Ford solid-state ignition system.

Introduction to Automotive Solid-State Electronics

1975 SOLID STATE

	Test Voltage Between	Should Be	If Not
Key On	Coil BAT terminal and ground (Module connected, DEC terminal grounded)	4.9 to 7.9 volts	Low reading - Check primary wiring. High reading - Replace resistance wire
	Socket 4 (Red) and ground	Battery voltage ± 0.1 volt	Repair red wire, check connectors
	Socket 1 (Green) and ground	Battery voltage ± 0.1 volt	Check green wire to coil, check coil
Cranking	Socket 5 (White) and ground	8 to 12 volts	Repair white wire, check connectors
	Socket 6 (Blue) and ground (Jump socket 1 to 8)	more than 6 volts	Check coil connections, check bypass circuit
	Socket 3 (Orange) and socket 7 (Purple)	0.5 volt ac or any dc voltage	Replace magnetic pickup (stator)
	Test Resistance Between	**Should Be**	**If Not**
Key Off	Socket 3 (Orange) and Socket 7 (Purple)	400 to 800 ohms	Replace magnetic pickup (stator) or repair ground connection
	Socket 8 (Black) and ground	0 ohms	
	Socket 3 and ground	more than 70,000 ohms	
	Socket 7 and ground	more than 70,000 ohms	
	Socket 4 (Red) and coil tower	7,000 to 13,000 ohms	Replace coil
	Socket 1 (Green) and Socket 6 (Blue)	1.0 to 2.0 ohms	
	Socket 1 (Green) and ground	more than 4.0 ohms	Check for short at coil DEC terminal or in wiring to DEC terminal
	Socket 4 (Red) and Socket 6 (Blue)	1.0 to 2.0 ohms	Replace resistance wire

1977-78 DURA-SPARK I

	Test Voltage Between	Should Be	If Not
Key On	Socket 4 (Red) and ground	Battery voltage ± 0.1 volt	Repair red wire, check connectors
	Socket 1 (Green) and ground	Battery voltage ± 0.1 volt	Check green wire to coil, check coil
	Socket 5 (White) and ground	8 to 12 volts	Repair white wire, check connectors
Cranking	Coil BAT terminal and ground (Jumper socket 1 to 8)— less than 30 seconds	more than 6 volts	Check coil connections, check bypass circuit
	Socket 3 (Orange) and 7 (Purple)	0.5 volt ac or any dc voltage	Replace magnetic pickup (stator)
	Test Resistance Between	**Should Be**	**If Not**
Key Off	Socket 3 (Orange) and Socket 7 (Purple)	400 to 800 ohms	Replace magnetic pickup (stator) or repair ground connection
	Socket 8 (Black) and ground	0 ohms	
	Socket 3 and ground	more than 70,000 ohms	
	Socket 7 and ground	more than 70,000 ohms	
	Socket 4 (Red) and coil tower	7,000 to 13,000 ohms	Replace coil
	Coil BAT and DEC terminal	0.5 to 1.5 ohms	
	Socket 1 (Green) and ground (2-wire connector connected)	more than 4 ohms	Check for short at coil DEC terminal or in wiring to DEC terminal

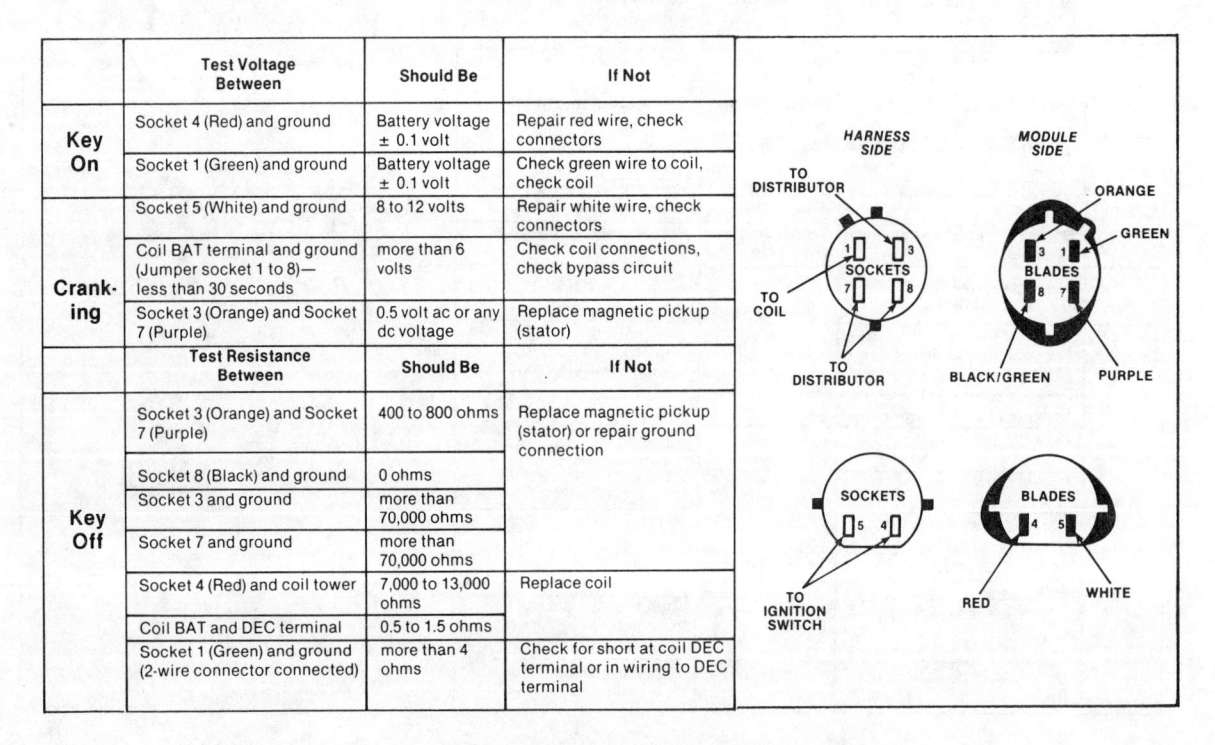

Fig. 4–14. Ford solid-state ignition system. *(Cont'd)*

Ignition Systems

	Test Voltage Between	Should Be	If Not
Key On	Coil BAT terminal and ground (Module connected DEC terminal grounded)	4.9 to 7.9 volts *6.0 volts ± .5	Low reading - Check primary wiring. High reading - Replace resistance wire
	Socket 4 (Red) and ground	Battery voltage ± 0.1 volt (* ± .2)	Repair red wire, check connectors
	Socket 1 (Green) and ground	Battery voltage ± 0.1 volt	Check green wire to coil, check coil
Cranking	Socket 5 (White) and ground	8 to 12 volts *Battery voltage	Repair white wire, check connectors
	Socket 3 (Orange) and Socket 7 (Purple)	0.5 volt ac or any dc voltage *Any fluctuation	Replace magnetic pickup (stator)
	Test Resistance Between	**Should Be**	**If Not**
Key Off	Socket 3 (Orange) and Socket 7 (Purple)	400 to 800 ohms	Replace magnetic pickup (stator) or repair ground connection
	Socket 8 (Black) and ground	0 ohms	
	Socket 3 and ground	more than 70,000 ohms	
	Socket 7 and ground	more than 70,000 ohms	
	Socket 4 (Red) and coil tower	7,000 to 13,000 ohms	Replace coil
	Coil BAT and DEC terminal	1.0 to 2.0 ohms	
	Socket 1 (Green) and Socket 4 (Red)	3.0 to 4.0 ohms- 1976	Check for short at coil DEC terminal or in wiring to DEC terminal
		1.7 to 3.7 ohms- 1977 (*78 SSI)	
	Socket 4 (Red) and coil BAT terminal	1.0 to 2.0 ohms- 1976	Replace resistance wire
		0.7 to 1.7 ohms- 1977 *78 1.35 ohms	

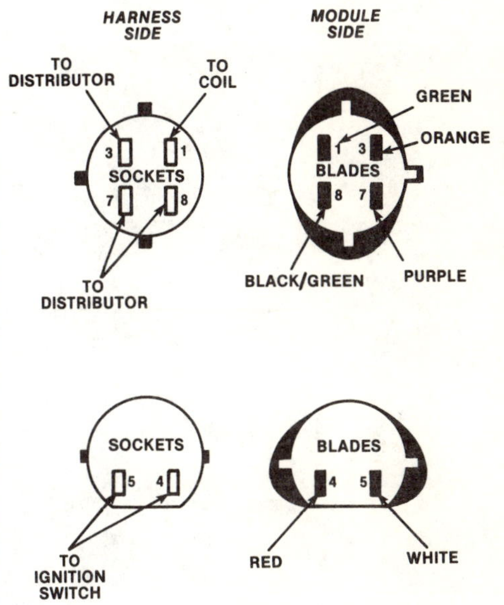

1978 American Motors SSI Specs — All Others Same as Ford

Courtesy Champion Spark Plug Company

Fig. 4-14. Ford solid-state ignition system. *(Cont'd)*

43

CHAPTER 5

Emission-Control Systems

The word smog has been part of our vocabulary for many years now. The expression was coined from the words smoke and fog because it was originally thought that these were the main ingredients. It has since been proven that these are not the culprits. Actually, there are two types of smog. London-type smog is caused chiefly by sulfur dioxide fumes from industrial processes. Automobiles produce virtually no sulfur dioxide and so are not a factor in London-type smog. The Los Angeles-type smog, however, is caused by a photochemical reaction involving certain hydrocarbons, oxides of nitrogen, and sunlight. When atmospheric conditions are just right, these ingredients combine to form new and unpleasant compounds. The automobile, along with industry, is equally guilty of creating this second type of smog.

COMBUSTION IN AN AUTOMOTIVE ENGINE

Engineering textbooks describe the way an internal combustion engine burns gasoline by the following formula:

$$C_8H_{18} + O_2 \rightarrow H_2O + CO_2 + \text{heat}$$

The gasoline (C_8H_{18}, or octane) combines perfectly with the oxygen (O_2) in the air to form water vapor (H_2O), carbon dioxide (CO_2), and heat (the energy that powers the vehicle). Theoretically, no other noxious compounds are formed. Unfortunately, engines do not operate under such theoretically perfect conditions.

A real-world engine does produce, as a residual byproduct, some of the very ingredients that lead to the formation of Los Angeles-type smog. Even if the engine is receiving a *stoichiometric mixture*, that theoretically perfect ratio of fuel and air, it may not achieve perfect or complete combustion (burning) of all the fuel molecules. Some of these molecules, principally hydrocarbon molecules, pass through the engine unburned and come out the exhaust as an unwanted pollutant.

Even though the temperatures of combustion are extremely high, some of the hydrocarbons do not get burned, or oxidized as the engineers say. The unburned quantity is small and is measured in parts per million (ppm). But even this small amount, from enough engines, can cause a serious air-pollution problem. The molecules that escape burning are mainly those along the relatively cool walls of the combustion chamber. With careful engine design, this can be kept to a minimum.

Byproducts of Incomplete Combustion

Let's take a look at the term stoichiometric mixture. This is the mixture of fuel and air (actually the oxygen in the air) that should result in nearly perfect combustion and the ratio that the oxygen sensor, to be described in the following chapters, attempts to maintain through the action of the on-board computer. The mixture is about 14.7 parts air (by weight) to one part fuel (by weight). At this ratio, the amount of carbon monoxide produced is a minimum: in theory, it should not be produced at all but in practice there will always be some.

Carbon monoxide is a product of incomplete combustion. If all the hydrocarbon molecules of the gasoline could combine with all the oxygen molecules, we would end up with carbon dioxide, a rather benign byproduct. But since such perfection does not exist in the internal combustion engine, we get traces of carbon monoxide, a rather undesirable byproduct. Fig. 5-1 shows the relative amounts of carbon monoxide and unburned hydrocarbons (HC) for various air–fuel ratios.

Generally, as the mixture becomes richer than stoichiometric, the quantities of carbon monoxide and unburned hydrocarbons go up. However, certain engine-operating conditions, such as acceleration, warmup, and heavy-load operation, require richer-than-stoichiometric mixtures. In the past, idling required rich mixtures to keep the engine running smoothly at low speeds but, today, engines idle close to the stoichiometric point.

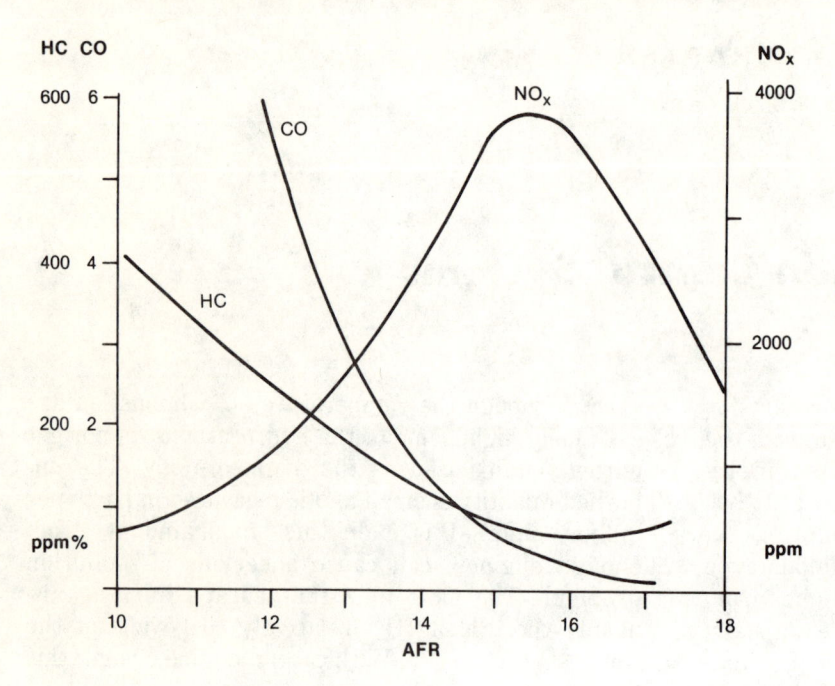

Introduction to Automotive Solid-State Electronics

Fig. 5-1. Relative amounts of exhaust emissions from an uncontrolled engine for various air-fuel ratios (AFR).

Formation of Oxides of Nitrogen

One of the big headaches with emission-control work is that as soon as you think you have solved one problem, another has been created. While lean or stoichiometric mixtures reduce CO and HC, they raise combustion temperatures. In turn, this higher temperature leads to the formation of oxides of nitrogen, another byproduct that contributes to Los Angeles-type smog. This can be seen in Fig. 5-1. There are several oxides of nitrogen, but we deal with them collectively and designate them as NO_x.

Here is what happens when enough NO_x, under the right conditions, gets into the atmosphere. Sunlight acting on NO_2 (nitrogen dioxide, a member of the NO_x family) releases atomic oxygen (O) to form a new type of oxygen (O_3) known as ozone. It is a very active form of oxygen and its characteristic odor is known to us all. Chemically, it is a powerful oxidizer and readily combines with other substances. It hardens and cracks rubber and can combine with yet another oxide of nitrogen to form nitric acid. The eye irritant PAN (peroxyacetal nitrate) is still another undesirable byproduct of the NO_x family.

There are several techniques available to the engineer to reduce the formation of NO_x. One of these is *exhaust gas recirculation*, or EGR as it is generally called. In this method, small amounts of exhaust gas are allowed to enter the intake manifold and mix with the incoming air–fuel charge. The recirculated exhaust gases help keep down the peak combustion temperatures and thereby reduce the probability of producing NO_x. Another technique is retarded ignition timing during low-speed operation.

In summary, we can say that emission control is concerned with the removal or minimization of three types of exhaust byproducts: carbon monoxide, unburned hydrocarbons, and oxides of nitrogen. Although lead compounds have been removed from gasoline, it is not because they present a primary form of exhaust pollution. Rather, they have been removed because of the adverse effect they have on catalytic converters and on the recently introduced oxygen sensors. We will be discussing these antipollution devices a little later.

POSITIVE CRANKCASE VENTILATION

Positive crankcase ventilation was the first approach to emission control and was employed on a fairly wide-scale basis starting in 1968. It had first been proposed some 30 years earlier as a means of purging the crankcase of vapors harmful to the engine. However, when it had been determined that about 20% of the hydrocarbons emitted by the engine were due to crankcase blowby, positive crankcase ventilation (PCV) became an effective method of emission control. Prior to adoption of PCV as an emission-control device, crankcase vapors were exhausted directly into the atmosphere by means of a "draft tube" extending from the crankcase. The forward motion of the vehicle created a draft that pulled these vapors out of the engine.

Emission-Control Systems

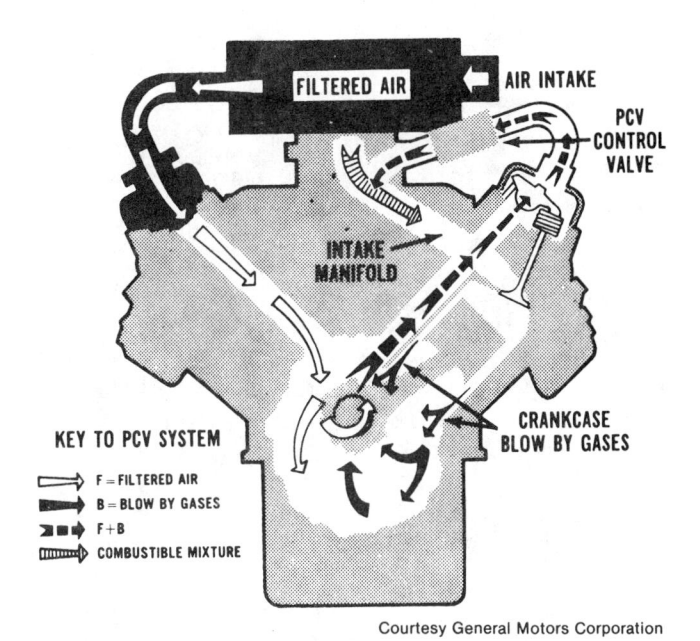

Fig. 5-2. Cross section of an engine utilizing the PCV system.

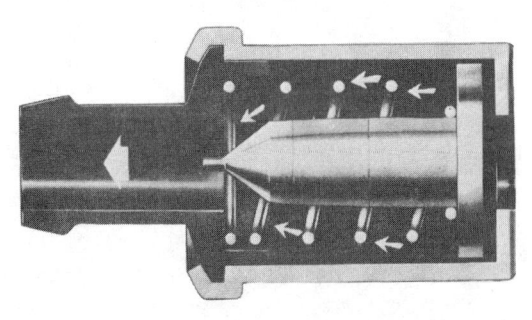

Fig. 5-3. Cross section of a typical PCV valve.

The modern PCV system (see Fig. 5-2) disposes of crankcase vapors by reintroducing them into the intake manifold where they are subsequently burned in the normal combustion process. As you can see, the system depends on intake manifold vacuum to draw purging air through the crankcase. This air flow picks up the blowby vapors escaping past the piston rings. The heart of the system is the PCV valve.

As shown in Fig. 5-3, this valve is spring loaded in order to limit the flow of purging air during periods of high manifold vacuum, such as at idle. A high vacuum causes the spring-loaded plunger to almost seal off the flow of air. This helps preserve the proper idle mixture adjustments. As manifold vacuum drops under higher speeds or heavier loads, the spring unseats the plunger, allowing a greater air flow and thus effective purging of the crankcase.

Most vehicle maintenance recommendations advise replacing the PCV valve at regular intervals, usually every 12,000 miles. If it should become stuck due to carbon or other internal buildup, it could upset the proper idle mixture or prevent proper purging of the crankcase. A common test for sticky PCV valves is to shake them while listening for the click of the plunger. If no clicking sound is heard, a stuck valve is indicated.

EVAPORATION CONTROL SYSTEMS

Another source of unburned hydrocarbons (approximately 15% of the total) comes from the vehicle's fuel tank and carburetor bowl. This is the result of normal evaporation of fuel through the gas-tank filler cap and the carburetor air horn. The methods and devices used to handle these emissions are known as *evaporation control systems*.

A typical system is shown in Fig. 5-4. The heart of the system is the charcoal canister, which serves as a temporary storage chamber for the fuel vapors when the engine is not running. Charcoal has the ability to absorb many times its own weight of vapors. To prevent any liquid fuel from entering the canister, a vapor separator is used. Should any fuel enter the lines, it is collected here and returned to the fuel tank.

A purge valve (see Fig. 5-4) permits the charcoal canister to be cleared of accumulated vapors after the engine is started. This valve connects the canister to the intake manifold, with the flow of purge air being regulated by manifold vacuum. Note the filter in the bottom of the canister; it is normally replaced about once a year or every 12,000 miles.

A special tank filler cap must be used with these systems. See Fig. 5-5. It is a two-step design similar to those used on pressurized radiator systems. Its purpose is twofold: (1) to prevent fuel from gushing out when the cap is removed and the tank is slightly pressurized and (2) to prevent tank damage in case a system failure causes either excessive tank pressure or vacuum. The locations of the major elements of this system are shown in Fig. 5-6. Note the roll-over valve that prevents raw fuel from leaking into the canister in the event of vehicle roll-over.

AIR-INJECTION PUMP

A small percentage of the hydrocarbon fuel molecules manage to pass through the engine without being burned (oxidized). One method of correcting this situation is to burn them *after* they leave the combustion chamber. This can be done by adding additional oxygen, in the form of air, to the exhaust

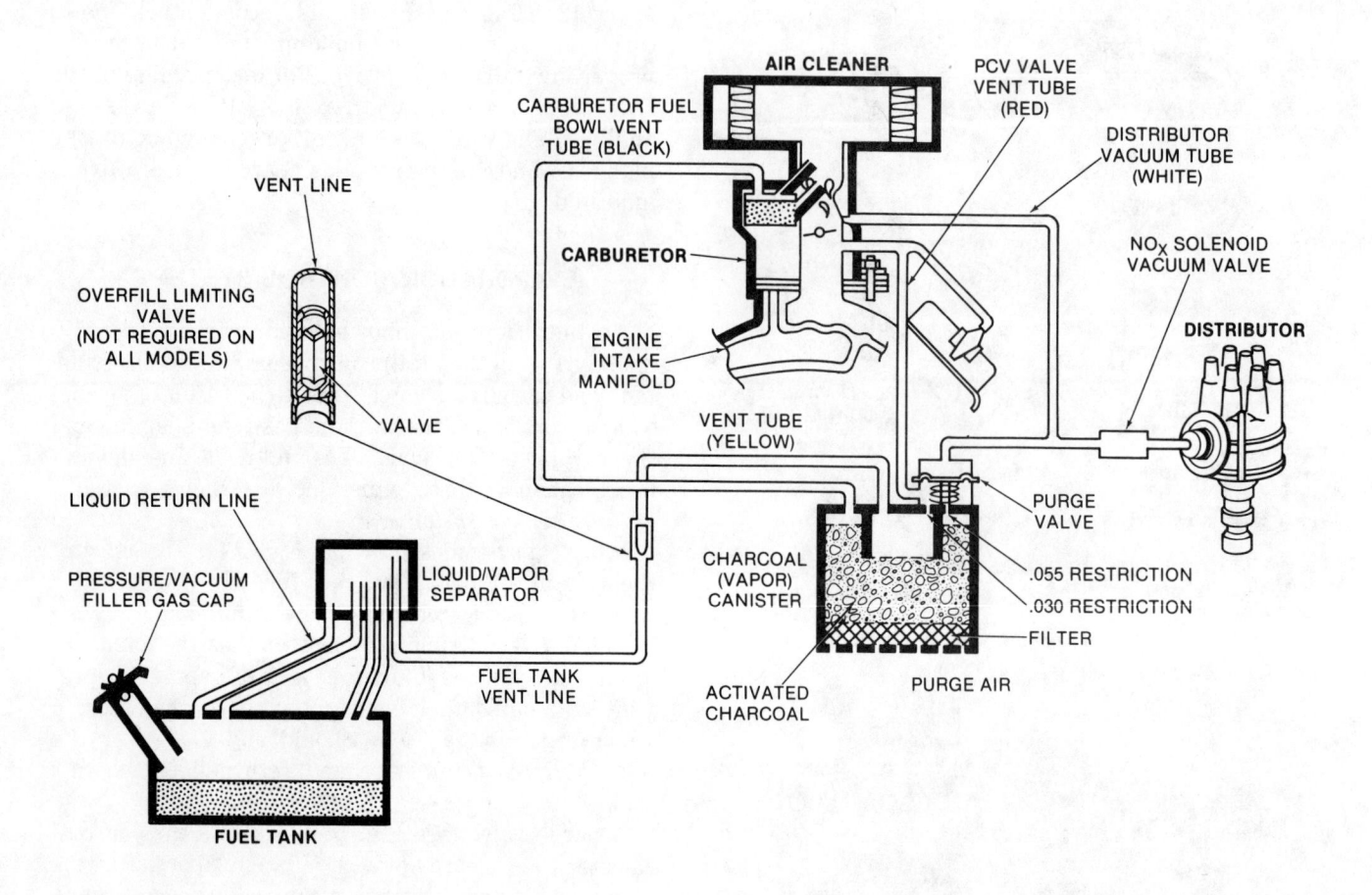

Introduction to Automotive Solid-State Electronics

Fig. 5-4. Typical evaporative emission-control system.

gases just as they enter the exhaust manifold. Still at a very high temperature from the combustion process, the as yet unburned hydrocarbons combine with the oxygen and exit the exhaust system as water vapor and carbon dioxide.

A typical system for accomplishing this task is shown in Fig. 5-7. The engine-driven air pump feeds low-pressure air directly into the exhaust manifold opposite each exhaust valve. Although only one exhaust port is shown, each cylinder's exhaust port receives air. The purpose of the diverter valve, shown in detail in Fig. 5-8, is to prevent backfires during deceleration. Normally, this valve is held open, allowing full air flow to the exhaust manifold, by a spring-loaded diaphragm connected to the intake manifold. When intake manifold vacuum suddenly increases during deceleration, the diaphragm spring compresses and closes off the air passage to the exhaust manifold. The air from the pump is diverted to the atmosphere. Because the diaphragm has a small bleed hole, this cutoff is only momentary.

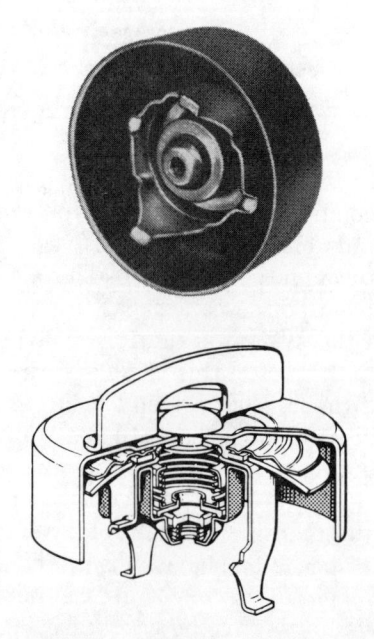

Fig. 5-5. Pressure-vacuum relief filler cap used with evaporative emission-control systems.

Emission-Control Systems

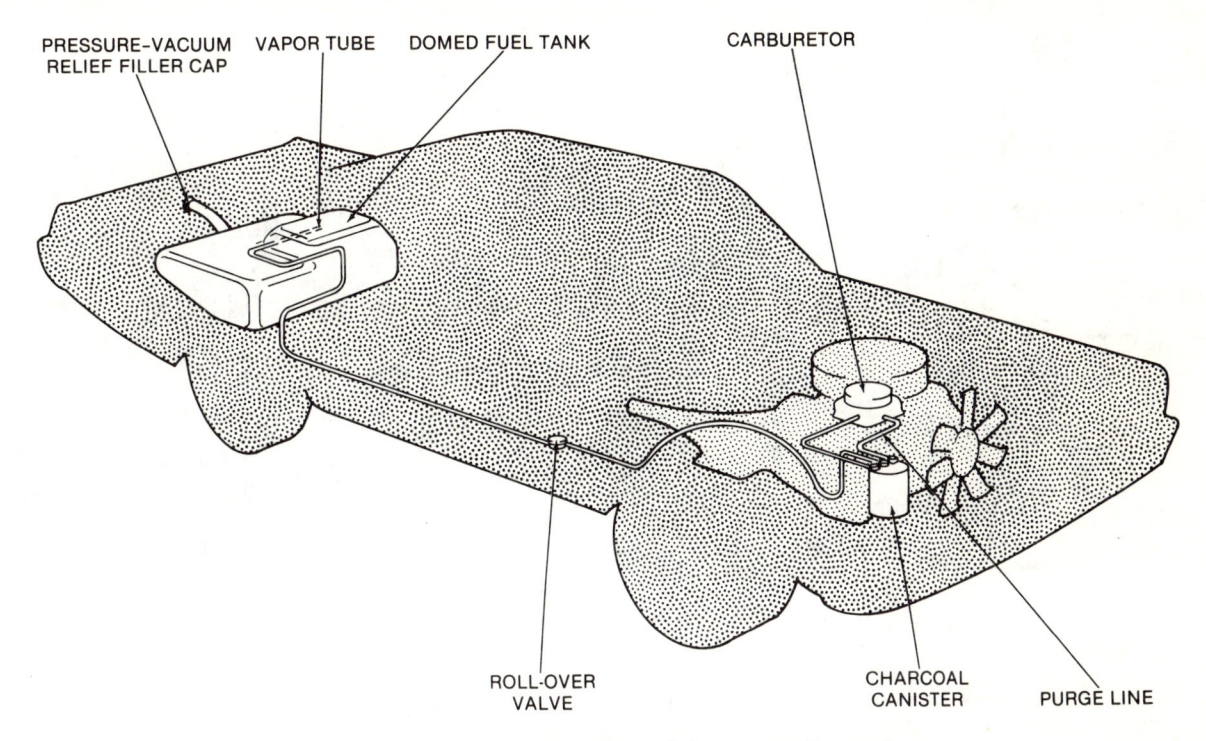

Fig. 5-6. Location of major parts of the evaporative system.

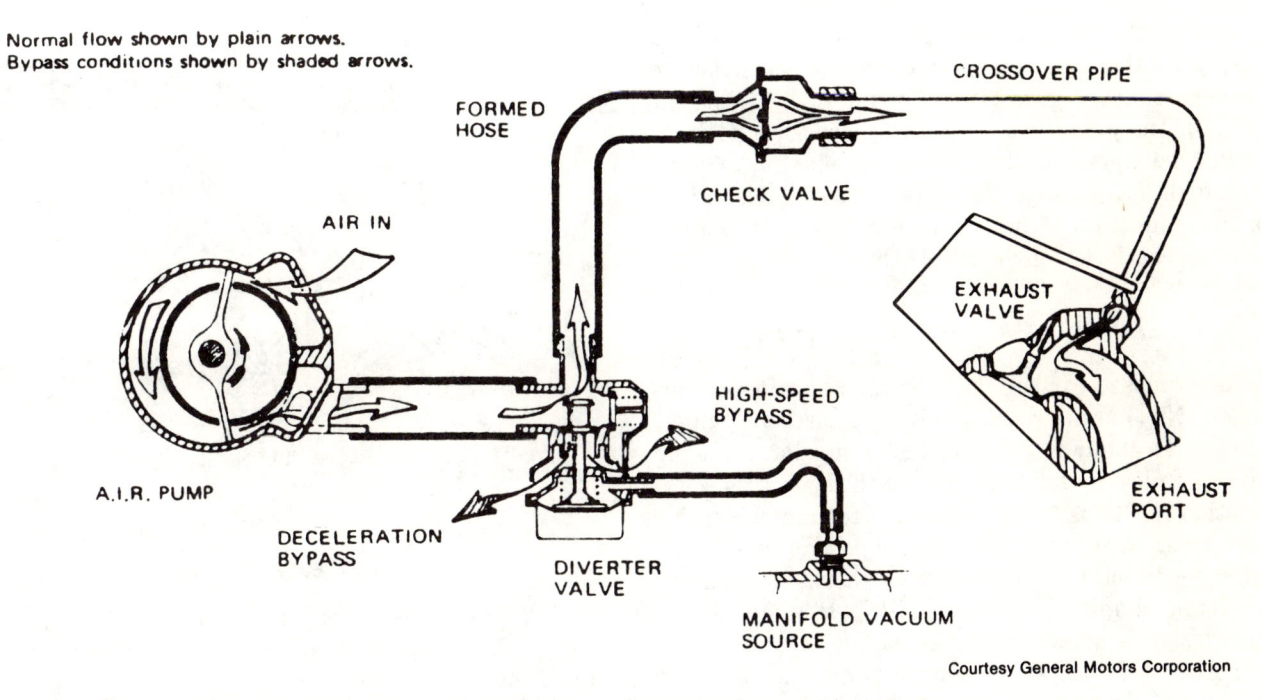

Fig. 5-7. Schematic of a typical air-injector reactor system.

Courtesy General Motors Corporation

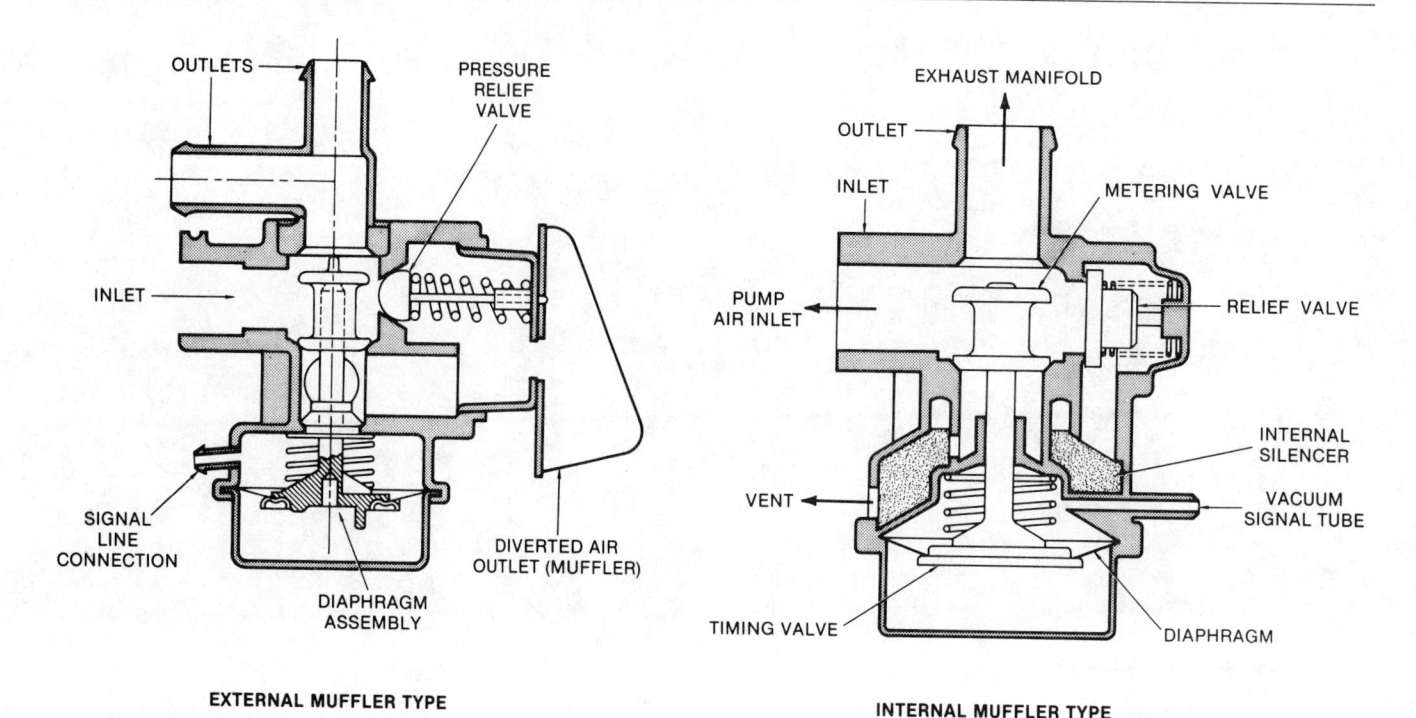

Introduction to Automotive Solid-State Electronics

Fig. 5-8. Cross section of a diverter valve.

In case the pump builds up too high a pressure, a pressure-relief valve opens and vents the air. Also, to prevent exhaust gases from flowing back to the air pump if the pump fails, a one-way check valve is installed between the diverter valve and the exhaust manifold. The air-injector system may be used alone or in conjunction with other emission-control systems such as catalytic converters or thermal reactors.

EXHAUST-GAS RECIRCULATION

As discussed earlier, the formation of oxides of nitrogen (NO_x) is the result of high combustion temperatures. It follows that one technique for reducing NO_x is to lower this temperature: Exhaust-gas recirculation (EGR) is the method used to accomplish this objective. In its simplest form, it is merely a small passage directly connecting the exhaust manifold to the intake manifold. An early EGR system employing this design is shown in Fig. 5-9.

This system was working at all times and could not be cut off. In order to provide better engine operation during idling, warmup, and full-throttle conditions, most engines now use a controlled EGR system. Fig. 5-10 shows the most basic form of this system. By locating the control vacuum line above the throttle plates, the EGR valve will be closed during normal idle operation and provide better idling characteristics.

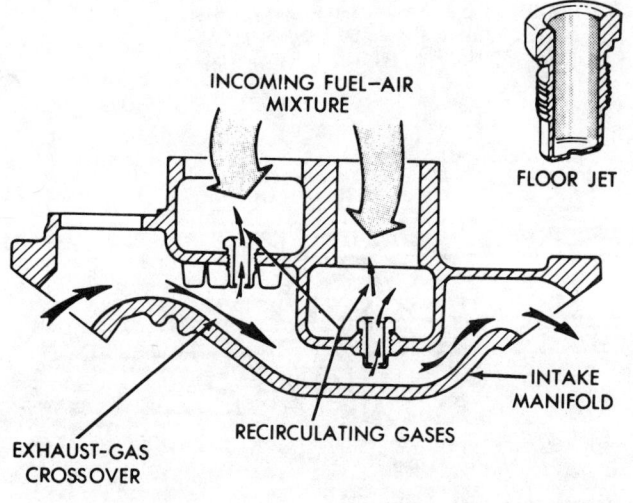

Fig. 5-9. Early EGR system.

Fig. 5-11 shows a cross section of an actual EGR valve. When the applied vacuum is low or nonexistent, the diaphragm spring causes the valve to close against the valve seat, closing off the flow of exhaust gas into the intake manifold. Various EGR vacuum-control systems, some purely mechanical and others electromechanical, are employed by the various car makers. Starting in 1980, some vehicles use computer-controlled EGR valves. These will be discussed in a later chapter. Next, some of the nonelectronic control methods are discussed.

50

Emission-Control Systems

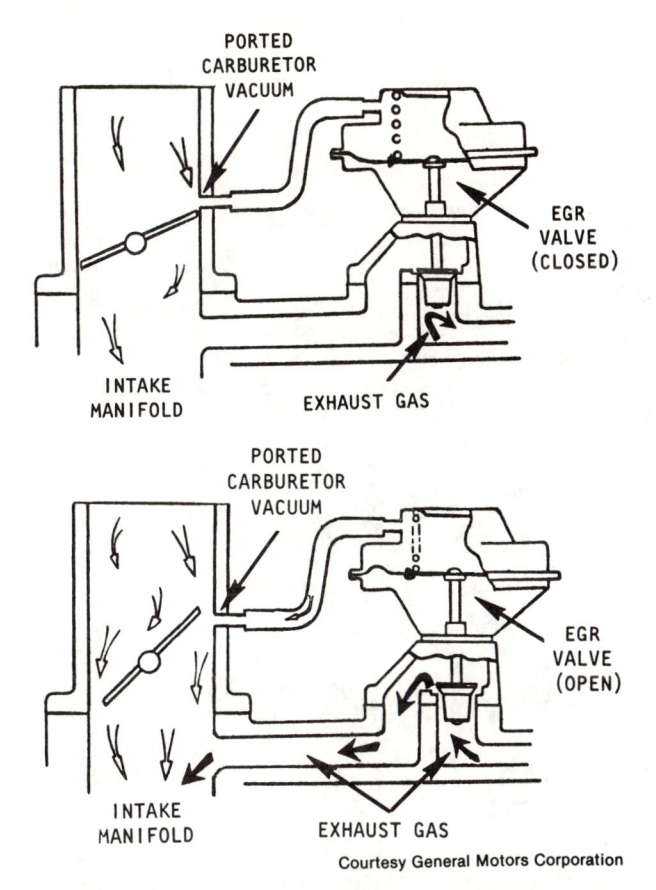

Fig. 5-10. Schematic view of a basic EGR system employing an EGR valve.

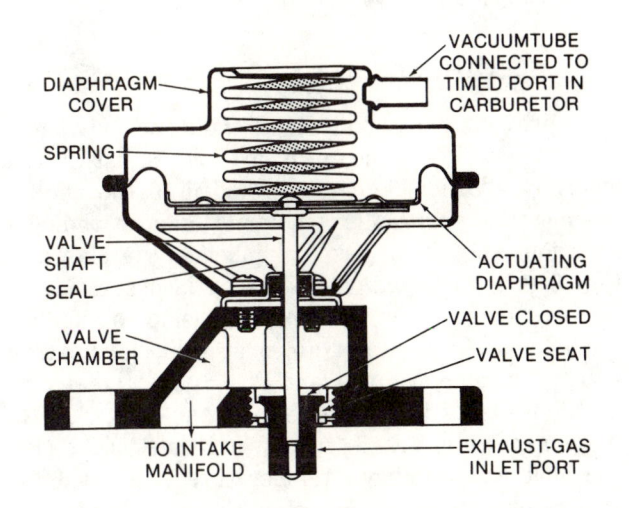

Fig. 5-11. Cross section of an actual EGR valve.

Dual-Diaphragm EGR Valve

To achieve better control over the recirculated exhaust gas, a dual-diaphragm EGR valve is sometimes used. See Fig. 5-12. Although it works in basically the same manner as the valve in Fig. 5-11, it is more responsive to changes in engine load. Note the combined action of exhaust gas and manifold vacuum.

Thermal Cutoff

It is desirable, from the standpoint of drivability, to prevent EGR operation during engine warmup. For this reason, most EGR systems have some type of temperature-sensitive vacuum-control valve to cut off EGR vacuum when the coolant temperature, or in some cases the underhood temperature, is below a certain level. A typical thermal cutoff valve is shown in Fig. 5-13. In addition, some systems have a time-delay control that prevents EGR operation for the first 30 seconds or so of engine operation. See Fig. 5-14. In those systems in which the EGR function is controlled by an on-board computer, EGR operation is determined by the program residing within the computer.

Venturi-Vacuum EGR

This system makes use of the carburetor venturi vacuum to control the EGR valve. But because of the rather weak vacuum signal available at the venturi, a vacuum amplifier must be used to supply enough vacuum to actuate the EGR valve. A typical system using venturi vacuum is illustrated in Fig. 5-15. Such a system makes the EGR function more responsive to changes in engine speed, but other than this, the system operates much the same as previously described systems.

As a rule, EGR systems require little maintenance other than to make sure the vacuum hoses are properly connected and are in good condition. In many systems, you can observe the action of the EGR valve or its actuating rod. One good way to do this is to set the engine speed at fast idle (engine at normal temperature, of course) and then disconnect the EGR vacuum hose. The speed should increase about 50 rpm or more (or whatever value is specified by the manufacturer) if the EGR valve is working properly. An EGR valve stuck in the open position during idle usually causes poor idling characteristics.

THERMOSTATICALLY CONTROLLED AIR CLEANER

Most engines produced since 1968 use a thermostatically controlled air cleaner to shorten the warmup period and improve drivability during this period. These systems allow air heated by the exhaust manifold to be drawn into the air cleaner during the

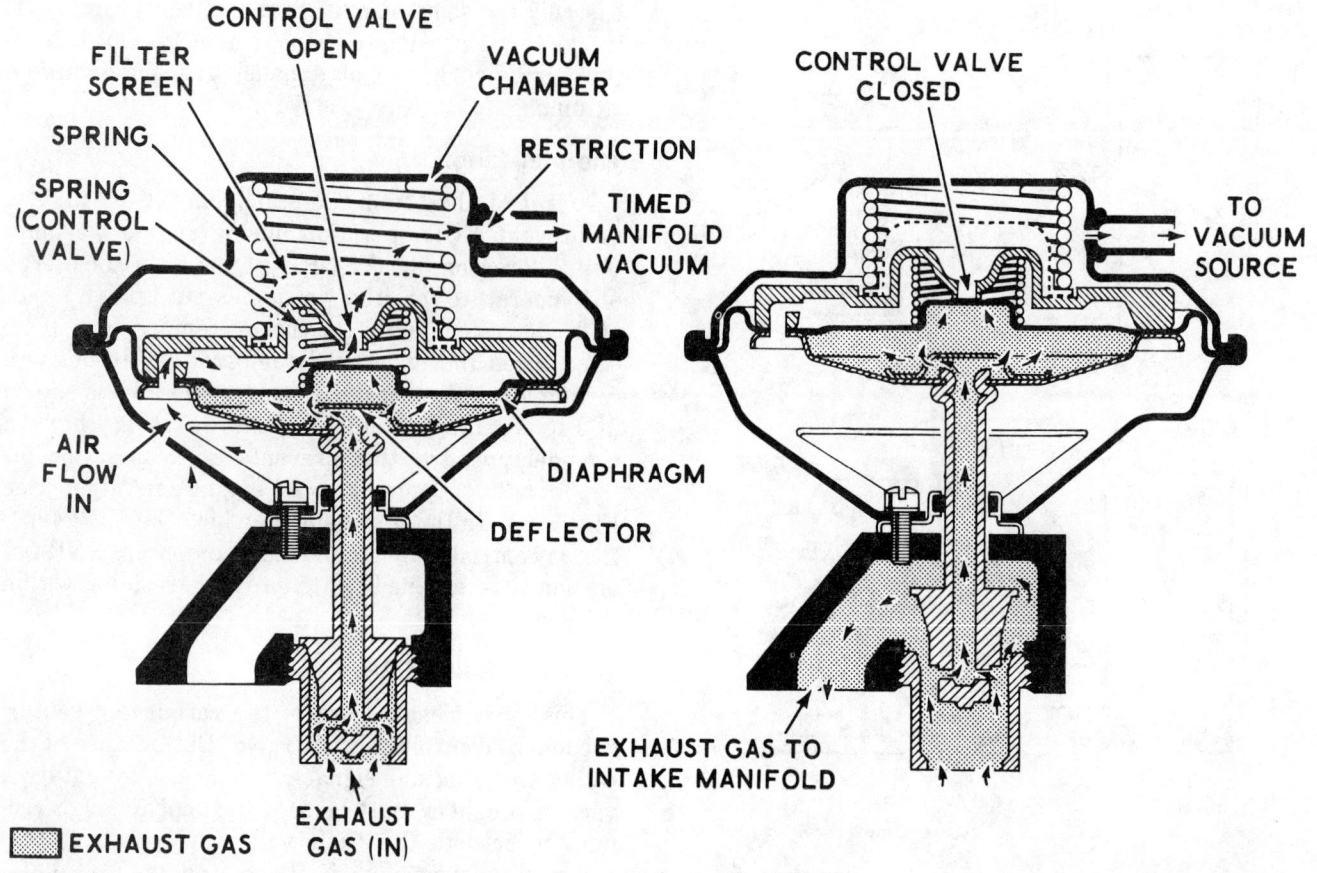

Introduction to Automotive Solid-State Electronics

Fig. 5-12. Cross section of a dual-diaphragm EGR valve.

Courtesy General Motors Corporation

warmup period. After operating temperatures are reached, underhood air is then drawn in. By doing this, not only is carburetor icing minimized, but the carburetor can be adjusted for leaner mixtures resulting in better fuel mileage and lower hydrocarbon emissions.

Two basic systems are used to accomplish this. The first is shown in Fig. 5-16. A thermal sensor controls the amount of vacuum applied to the vacuum diaphragm that controls the air-control valve. When the entering air temperature is below a certain level, the sensor allows full vacuum to the vacuum diaphragm, which then opens the control valve to admit heated air into the air cleaner. As the temperature of the incoming air rises, the control valve gradually closes until the heated air is cut off.

Another system is illustrated in Fig. 5-17. This system operates by direct mechanical action between the internal thermostat and the heat-control valve. Occasionally, a vacuum assist is added to the system, but the basic purpose of either system is the same. The complete range of operation is shown in Fig. 5-18.

ELECTRIC-ASSIST CHOKE SYSTEMS

To prevent excessively long choke periods, most engines now use some form of choke assist system whereby additional heat is supplied to the choke's thermostatic spring. This not only aids in reducing emissions, but improves fuel mileage. As shown in Fig. 5-19, this is accomplished by incorporating an electric heater within the choke housing to augment the normal flow of hot air past the thermostatic spring. In some systems (see Fig. 5-20), the electric heater provides all of the heat for the choke. In the system shown in Fig. 5-19, a switch cuts off the electric heat when ambient temperature is below 60°F. This allows normal choke action during cold weather operation.

Still another system is shown in Fig. 5-21. This method gives three modes of choke control. At engine temperatures below 58°F, the choke heater receives only partial heater current. At about 58°F, it receives full heater current and when the temperature is over 100°F, heater current is cut off. These various modes are controlled by the choke-control unit.

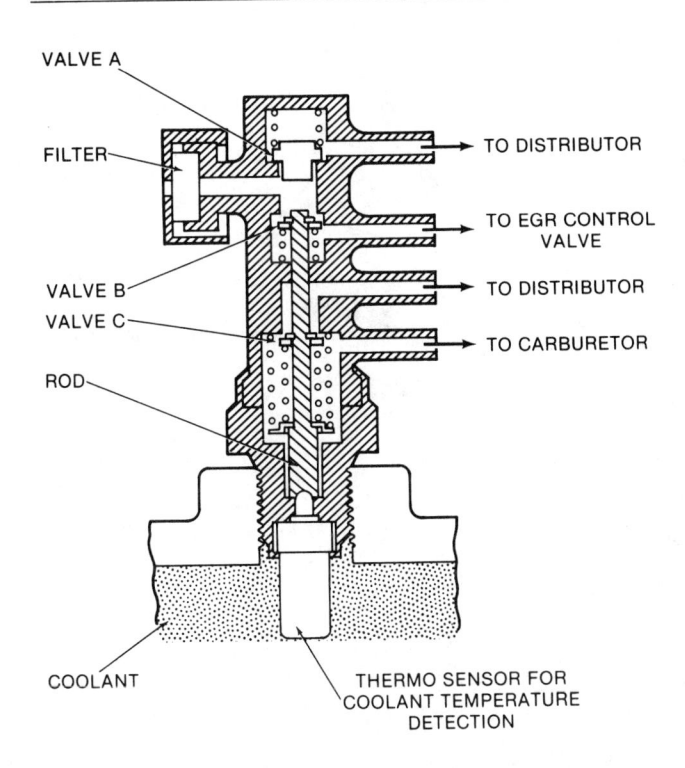

Emission-Control Systems

Fig. 5-13. Cross section of a typical EGR thermal cutoff valve.

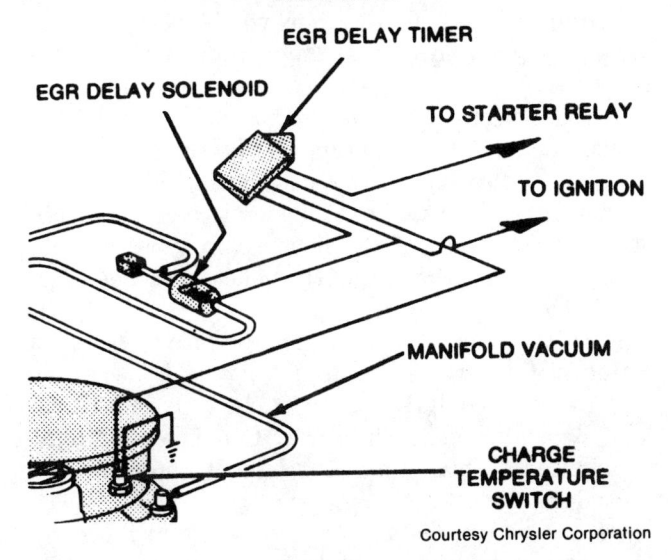

DESCRIPTION	WORKING TEMPERATURE
VALVE A (OVER TEMPERATURE PROTECTION VALVE)	OPENS AT ABOVE 203°F (95°C)
VALVE B (THERMOVALVE FOR EGR CONTROL)	CLOSES AT ABOVE 104°F (40°C)
VALVE C (THERMOVALVE FOR OSAC)	CLOSES AT ABOVE 104°F (40°C)

Fig. 5-14. EGR system with time delay. *Courtesy Chrysler Corporation*

IDLE STOP SOLENOID

Engine run-on or "dieseling" after the ignition is turned off is not an uncommon problem. One of the main causes is high idling speed made necessary by emission-control requirements. The purpose of the idle stop solenoid is to prevent this condition by providing two levels of idle speed. It maintains the normal hot idle speed while the engine is running, but drops it to a lower speed the moment the ignition is turned off.

Fig. 5-22 shows a typical idle stop solenoid attached to the throttle-control linkage of the carburetor. When the ignition is on, the solenoid is energized causing the plunger to extend and hold the throttle lever in a higher-speed position. When the ignition is turned off, the plunger retracts, thus lowering the idle speed. Both of these speeds are usually adjustable. Although specifications vary, there is generally about a 150 or 300 rpm differential between the two settings.

ORIFICE SPARK ADVANCE CONTROL (OSAC)

One way to reduce the formation of NO_x is to delay the application of vacuum advance to the distributor. A system that accomplishes this is called the Orifice Spark Advance Control System. See Fig. 5-23. The main component in this system is a control valve containing a small orifice. It is located between the distributor vacuum advance port on the carburetor and the vacuum advance chamber on the distributor. A typical OSAC valve is shown schematically in Fig. 5-24. Note that this valve incorporates a temperature-sensing unit that bypasses the spark-delay function when the air temperature is below 58°F. This permits normal, unrestricted vacuum to the distributor at temperatures below that level to aid drivability.

When air temperature is above 58°F, the vacuum applied to the distributor is delayed by approximately 20 seconds (varies with different vehicles). This system can be checked for proper operation by measuring the vacuum applied to the distributor when engine speed is increased from idle to approximately 2000 rpm. The vacuum should increase slowly over a period of about 20 seconds. If it increases rapidly, the OSAC valve is defective.

SPARK-DELAY VALVE

This device performs essentially the same function as the OSAC valve described above; i.e., it delays the application of the vacuum to the vacuum advance unit

53

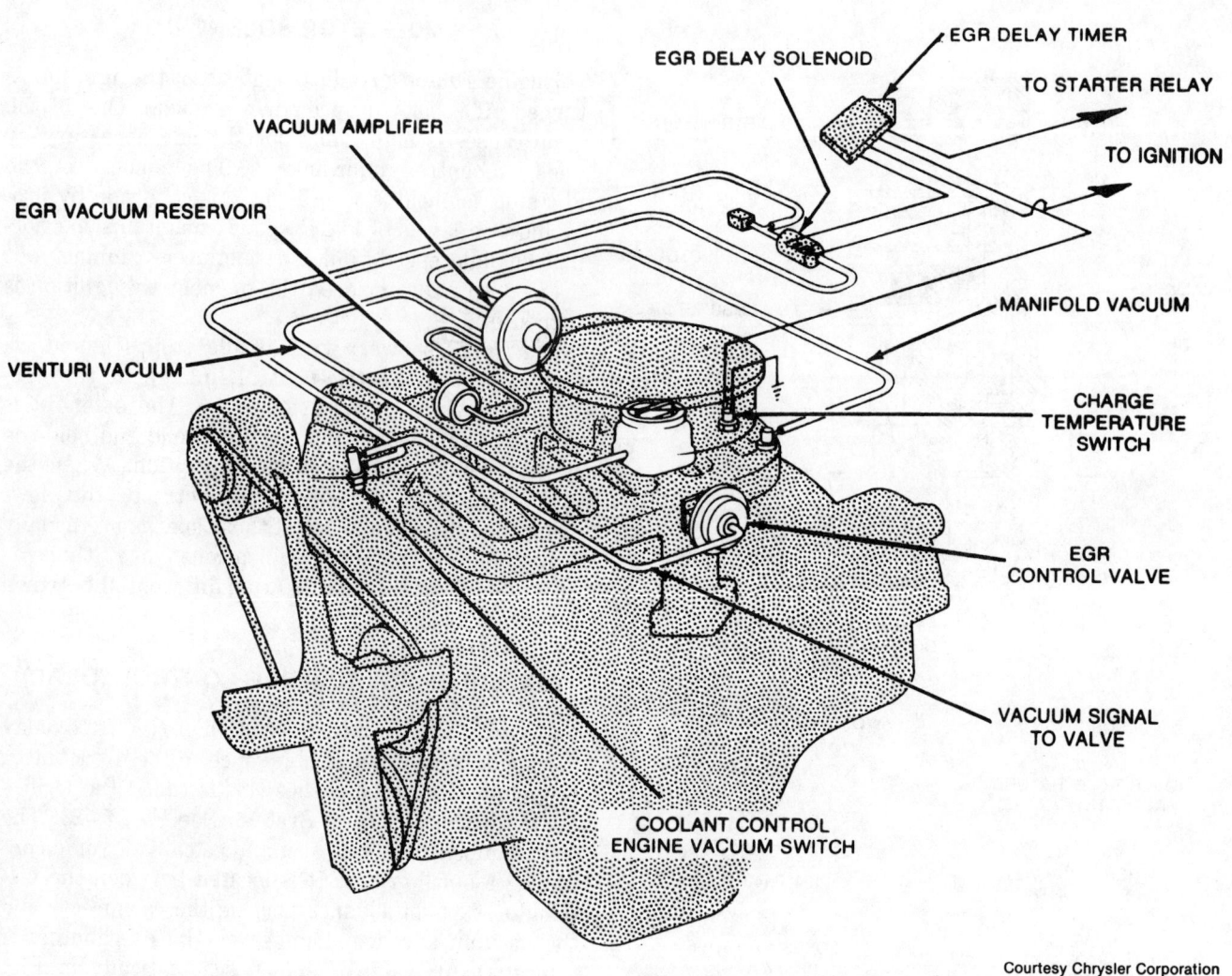

Introduction to Automotive Solid-State Electronics

Fig. 5-15. Schematic of a venturi-vacuum controlled EGR system.

Courtesy Chrysler Corporation

in the distributor. However, the restriction is one way, meaning that the distributor advances slowly but retards quickly. Fig. 5-25 shows a group of similar-looking vacuum-control valves. The one shown in the lower left is the type normally used for spark-delay purposes.

CATALYTIC CONVERTERS

Most of the emission-control systems and devices discussed so far have been designed to prevent or minimize the formation of undersirable byproducts such as HC, CO, and NO_x. The catalytic converter attacks the problem by oxidizing these byproducts after they have been formed.

A catalyst is a substance that promotes or accelerates a chemical reaction but does not enter into the reaction. It does not become a byproduct of that reaction. The principal catalytic materials used for automotive applications are platinum and, sometimes, palladium. These are used in what are known as oxidation-type converters; they promote the oxidation of HC and CO into H_2O (water vapor) and CO_2.

Two basic converter constructions are used: (1) the monolithic or honeycomb type and (2) the pellet type. Both types function in the same manner. The catalytic material is coated onto a one-piece ceramic substrate, in the case of the honeycomb type, or onto ceramic beads in the case of the pellet type. See Figs. 5-26 and 5-27.

A third type of catalytic converter, called a Three-Way Converter, is used where it is necessary to convert not only HC and CO but NO_x into harmless compounds. The Three-Way Converter contains, in addition to platinum, the element rhodium as a second catalytic agent.

Because of the catalytic converter (and more recently the oxygen sensors), leaded gasoline cannot be

Emission-Control Systems

used. Over a period of time, the lead compounds in the gas will coat the catalytic material and render it ineffective. Occasional use of leaded fuel, in emergency situations, is not considered harmful. The lead compounds that may have formed on the catalyst will tend to burn off if the use of unleaded fuel is quickly restored, provided the catalyst has not been made totally ineffective. Prolonged use of leaded fuel has been known to cause blockage in converters, with subsequent loss of engine performance. The effect is the same as a restricted exhaust system.

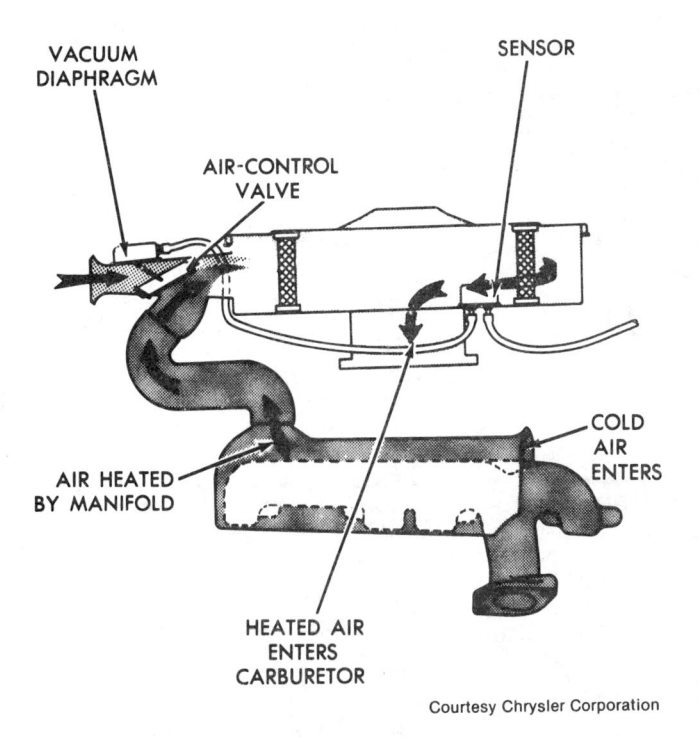

Fig. 5-16. Typical thermostatically controlled air-cleaner system.

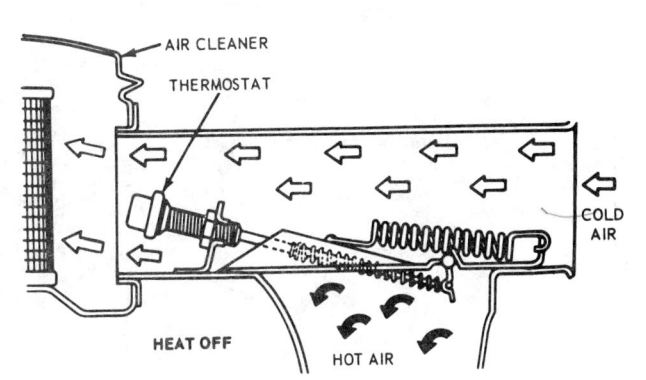

Fig. 5-17. Thermostatically controlled air-cleaner with air valve controlled directly by a thermostat.

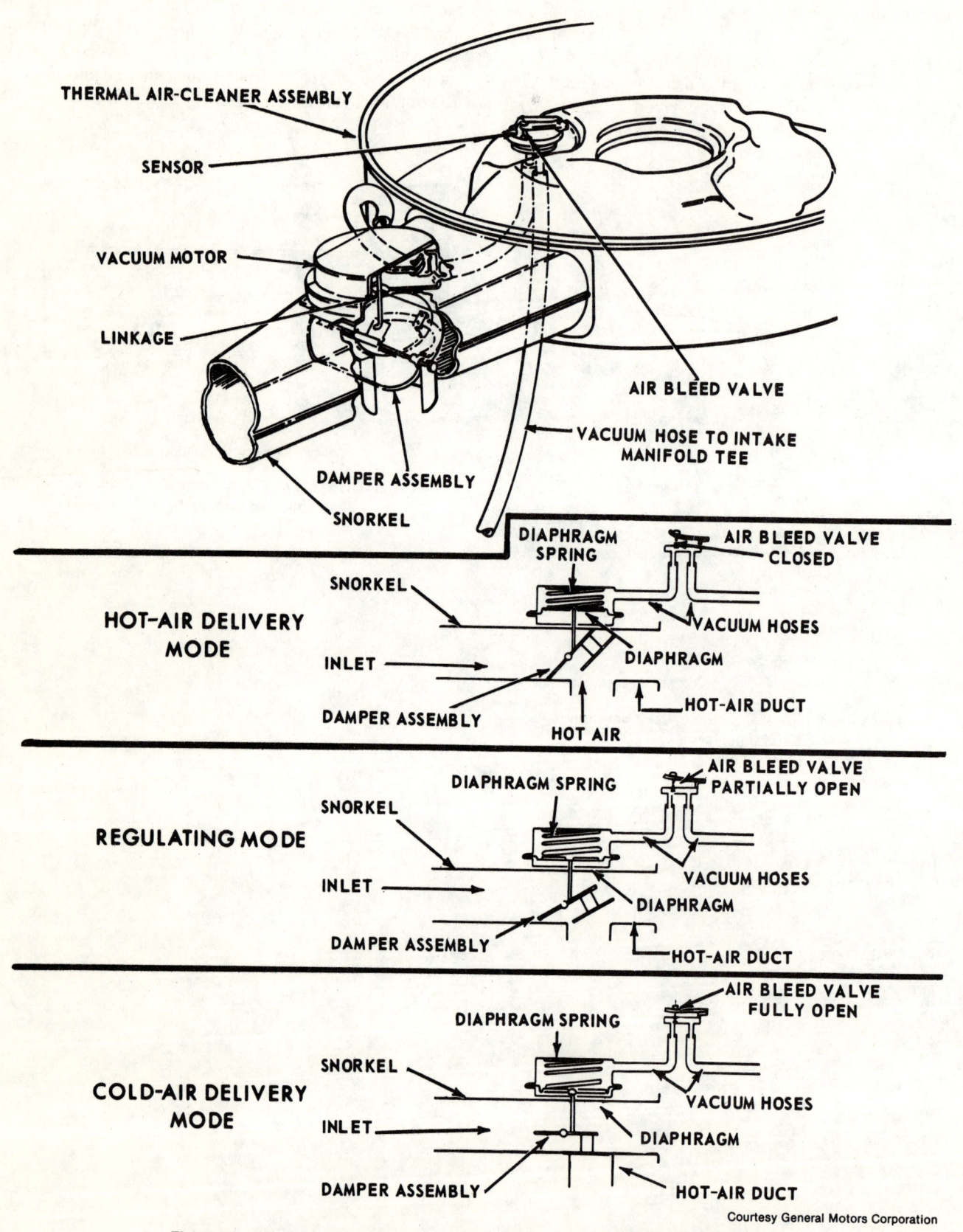

Introduction to Automotive Solid-State Electronics

Fig. 5-18. Various operating modes of the thermostatically controlled air cleaner.

Courtesy General Motors Corporation

Emission-Control Systems

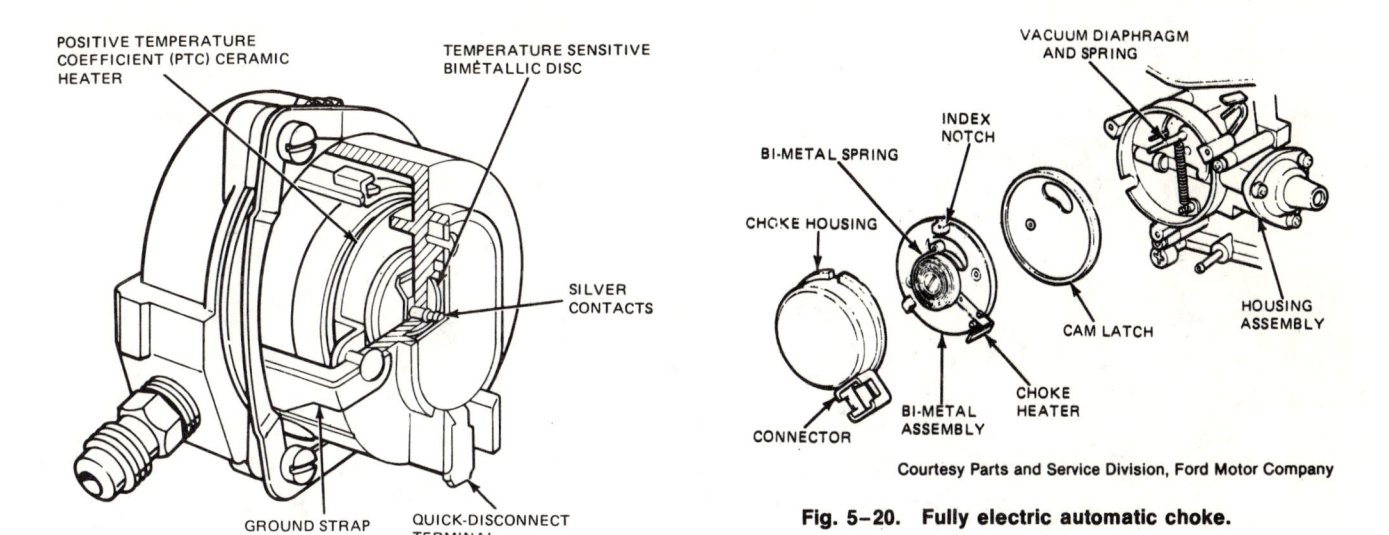

Fig. 5-19. Automatic choke with auxiliary electric heating. *Courtesy Parts and Service Division, Ford Motor Company*

Fig. 5-20. Fully electric automatic choke. *Courtesy Parts and Service Division, Ford Motor Company*

Fig. 5-21. Electric assist choke system with various levels of control. *Courtesy Chrysler Corporation*

57

Introduction to Automotive Solid-State Electronics

SOLENOID DE-ENERGIZED   **SOLENOID ENERGIZED**
(DISCONNECT IGNITION SWITCH LEAD AT BULLET CONNECTOR)

THROTTLE POSITION DETERMINED BY THROTTLE STOP SCREW

IDLE THROTTLE POSITION (RUNNING RPM) DETERMINED BY SOLENOID PLUNGER

LOCKNUT

TO IGNITION SWITCH

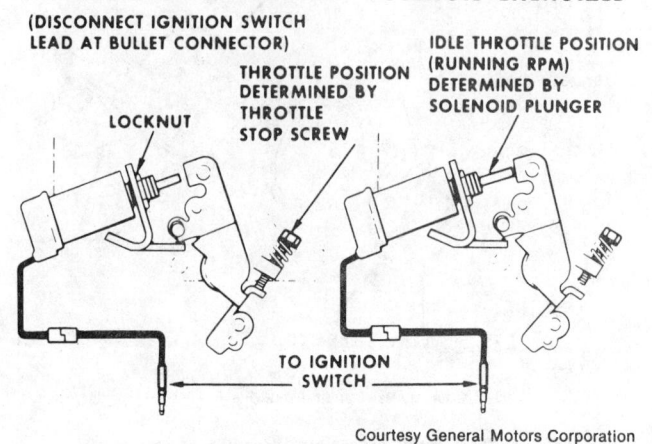

Courtesy General Motors Corporation

Fig. 5-22. Idle stop solenoid.

TO DISTRIBUTOR VACUUM ADVANCE UNIT

TEMPERATURE SENSING UNIT

TO CARBURETOR VACUUM

Courtesy Chrysler Corporation

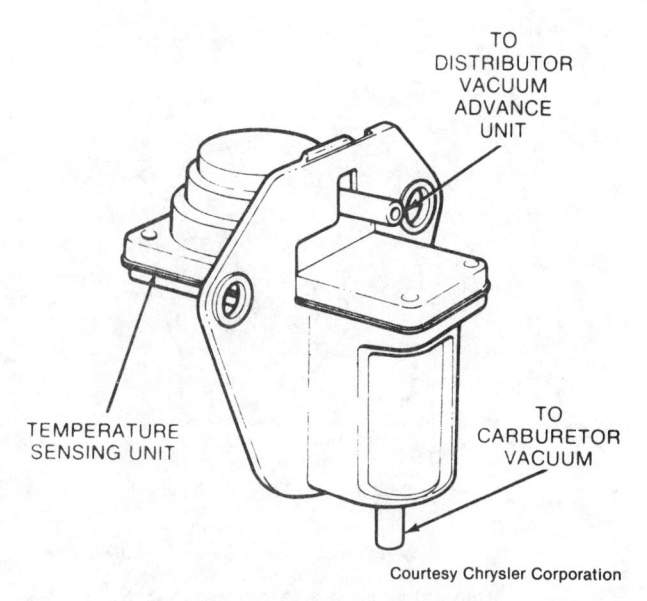

Fig. 5-24. OSAC valve assembly.

ORIFICE SPARK ADVANCE CONTROL VALVE (OSAC)

VACUUM LINES

CARBURETOR

DISTRIBUTOR

Courtesy Chrysler Corporation

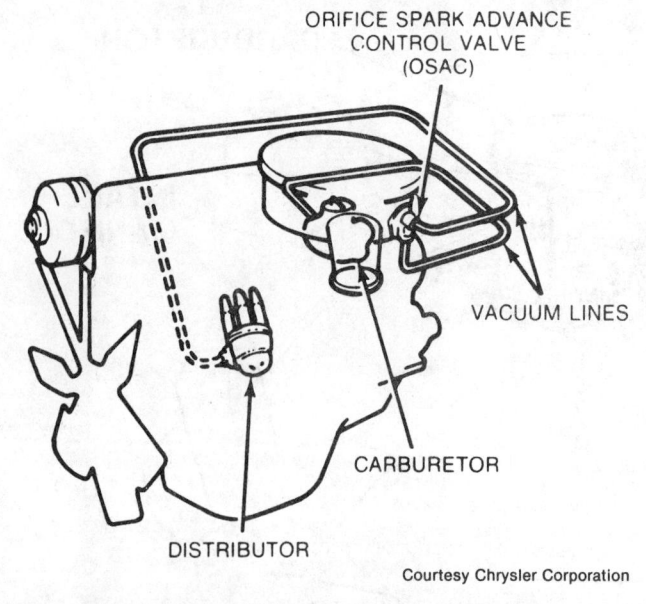

Fig. 5-23. Schematic of an Orifice Spark Advance Control.

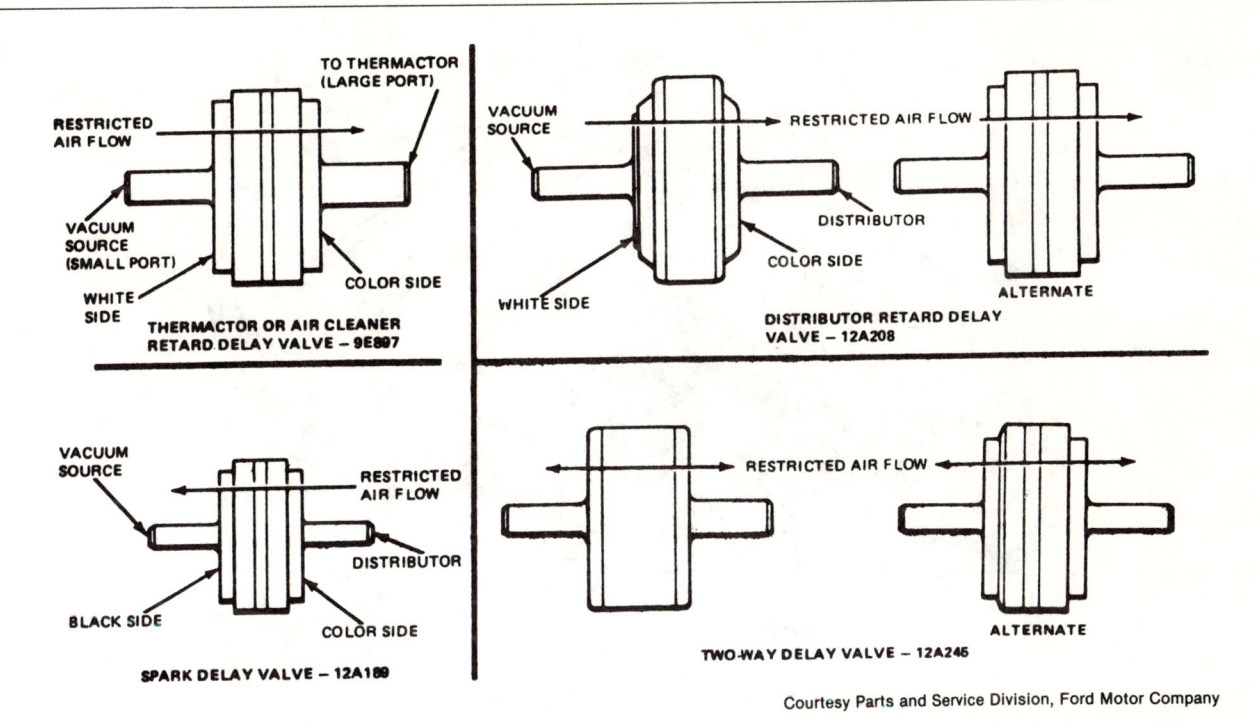

Fig. 5-25. Four types of vacuum delay valves.

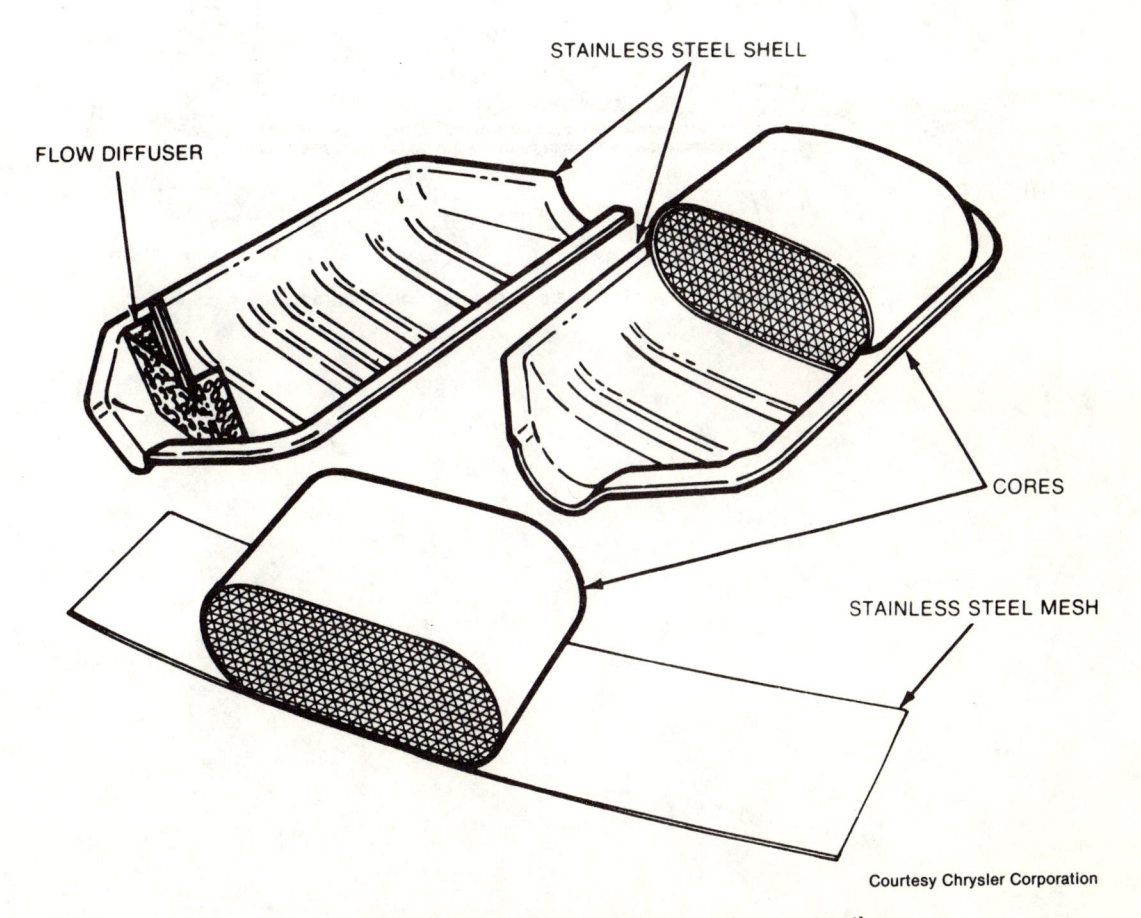

Fig. 5-26. Monolithic type catalytic converter construction.

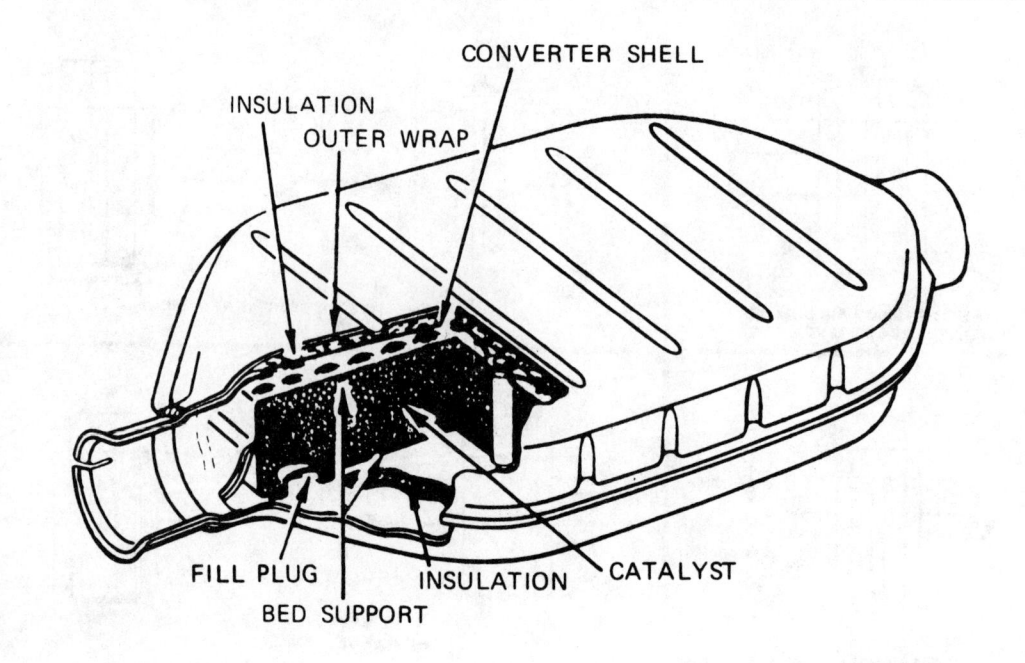

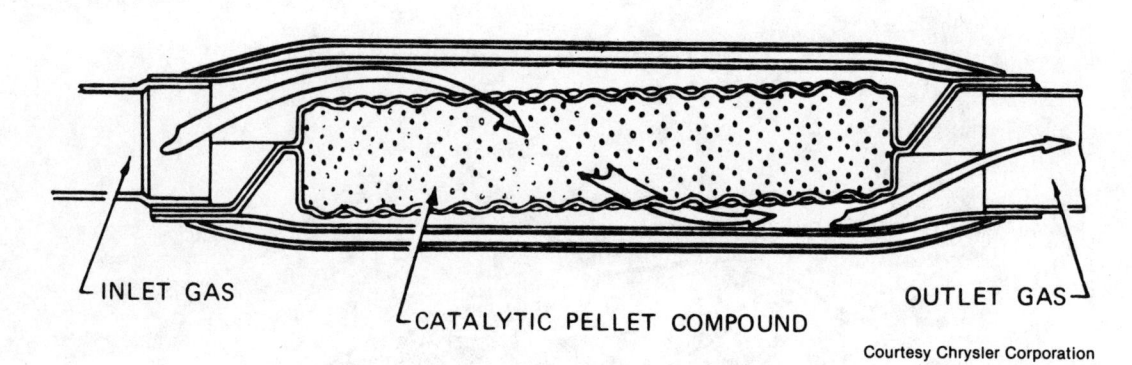

Introduction to Automotive Solid-State Electronics

Fig. 5-27. Pellet type catalytic converter construction.

Courtesy Chrysler Corporation

CHAPTER 6

Microprocessors, Computers, and Logic Systems for Automobiles

Until a few years ago, most engine control functions were handled by mechanical or electromechanical devices. In the days when exhaust emissions and fuel economy were not prime design requisites, such control devices were quite adequate. They were generally simple, easily understood, economical, and reasonably effective.

Unfortunately, they lacked the precision and the ability to respond to the variety of input conditions necessary to meet evermore stringent emission and fuel-economy requirements. To meet these requirements, today's engines demand an accuracy of control greater than we have ever known. This, in turn, has dictated the use of electronic, rather than mechanical, control over engine operation. As a result, we now have engines in which the basic functions of carburetion, spark timing, and emission control are handled, directly or indirectly, by solid-state logic systems and microcomputers.

The advent of large scale integration (LSI) has made possible on-board control systems that would have been impossible just a few years ago. In this chapter we will introduce you to the basic principles of these systems so that you can better understand their purpose and function.

LOGIC AND DECISION MAKING

One of the more vital roles of solid-state systems is their ability to perform logic functions and make decisions. Just what do we mean by *logic function*? It can be explained with a simple application.

We will start with some nonautomotive problems, since the principles are applicable to many situations. Assume that you are a plant engineer charged with the responsibility of making a punch press safer for the operators. Normally, the operator places a piece of raw metal between the punch press dies and then trips a foot pedal which causes the dies to close upon the metal. As long as the operator remembers to keep his hands clear of the dies, no accident will occur.

But accidents are just that—unplanned happenings. To prevent a situation where the operator may not remove his hands soon enough from the dies, you decide to replace the foot switch with two hand-operated switches wired in series with each other. See Fig. 6–1. Now the the operator must press both switches with his hands (they are placed far enough apart so they cannot be operated with one hand) before the punch press can go through its cycle.

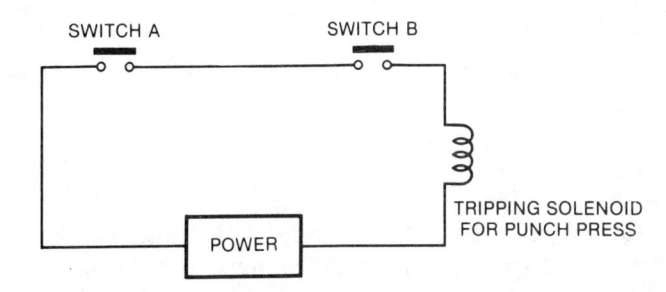

BOTH A AND B MUST CLOSE BEFORE TRIPPING SOLENOID ACTIVATES

Fig. 6–1. Basic AND circuit.

Both switch A AND switch B must be closed before the tripping mechanism can activate the press cycle. In logic terms this is known as an AND function. The AND must occur simultaneously before a signal (in this case, the current that activates the tripping mechanism) occurs.

Although the example we have just shown is an extremely simple AND function, it nevertheless is one of the basic logic functions. In solid-state systems, this function is performed by a device called an AND gate. See Fig. 6–2. Instead of mechanical switches, the AND gate employs transistor switches to achieve the same results. Only if a voltage is applied simultaneously to the A and B terminals does a voltage occur at the C, or output, terminal. If a voltage is present only at the A terminal but not the B terminal, or vice versa, no signal appears at the C terminal. Logic de-

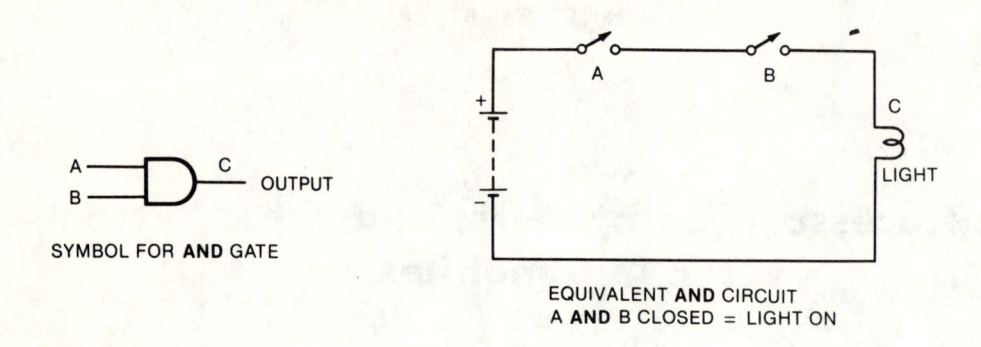

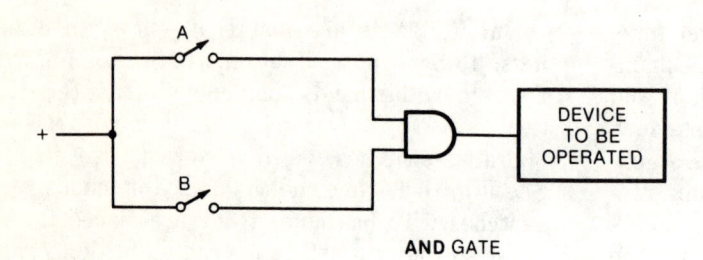

Introduction to Automotive Solid-State Electronics

SYMBOL FOR **AND** GATE

EQUIVALENT **AND** CIRCUIT
A **AND** B CLOSED = LIGHT ON

Fig. 6-2. Schematic representation of an AND circuit and how it responds to various input signals.

AND GATE

AND INPUT A B	GATE OUTPUT
0 0	0 (NO OUTPUT)
0 1	0 (NO OUTPUT)
1 0	0 (NO OUTPUT)
1 1	1 (OUTPUT)

signers use a shorthand notation to indicate whether a voltage is present or not. The number "1" means a voltage is present; the number "0" means no voltage is present.

Let's take another example that illustrates a different kind of logic function. Assume that you are a shopkeeper running the store by yourself. You must divide your time between the storeroom in the back and the salesroom up front. Customers can enter the shop through either of two doors. To inform you when a customer enters the shop when you are in the storeroom, you connect a switch to each door and wire them to a signal bell. The switches are wired in parallel so if either door opens, the bell will ring. In other words, if either door A OR door B opens, a signal develops. This is called an OR gate, as shown in Figs. 6-3 and 6-4. If a signal appears at either A OR B OR both, a signal also appears at C.

There is yet a third logic function called inversion (or the NOT function) that is performed by a solid-state device known as an *inverter* or NOT gate. See Fig. 6-5. If a signal (a "1") appears at the input of an inverter, no signal (a "0") appears at the output. The reverse is also true. With these three basic building blocks—the AND gate, the OR gate, and the NOT

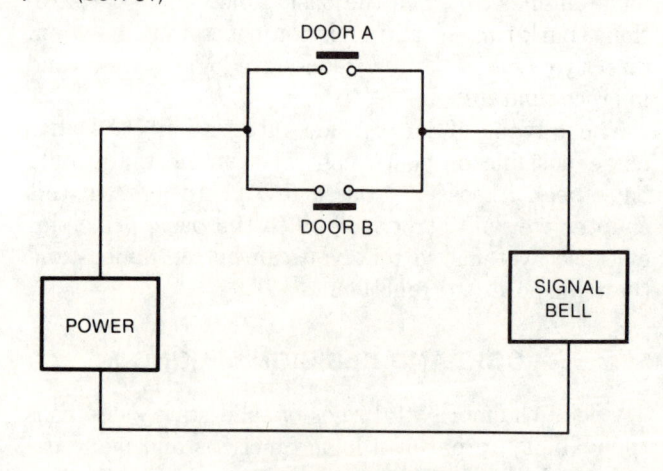

EITHER A OR B OR BOTH MUST CLOSE TO RING BELL

Fig. 6-3. Basic OR circuit

gate—a logic designer can create logic systems of tremendous complexity. These systems, in turn, make decisions based on the data fed into them. To see how a decision might be made by a logic system, consider the following example.

Again, you are a shopkeeper, but this time you have a clerk to help you. His main job is to tend the

62

Microprocessors, Computers, and Logic Systems for Automobiles

Fig. 6-4. Schematic representation of an OR circuit and how it responds to various input signals.

Fig. 6-5. Symbol for an inverter or NOT circuit and how it responds to input signals.

cash register and ring up sales. You must divide your time between the shop, your office, and the storeroom. The only time you want to be alerted is when someone enters the shop (remember, you have two entrances) and the clerk is NOT at the cash register. In other words, the signal bell should ring only if door A OR door B is opened AND the clerk is NOT at the register. Here is how a logic designer might put together the necessary logic modules to make this particular decision. See Fig. 6-6. Note that the system is made up of the basic AND, OR, and NOT gates.

Of course the logic units must be connected to suitable *transducers* or *input devices*. For the doors, these would be switches that send a voltage (a logic 1 state) to the OR gate when either door was opened.

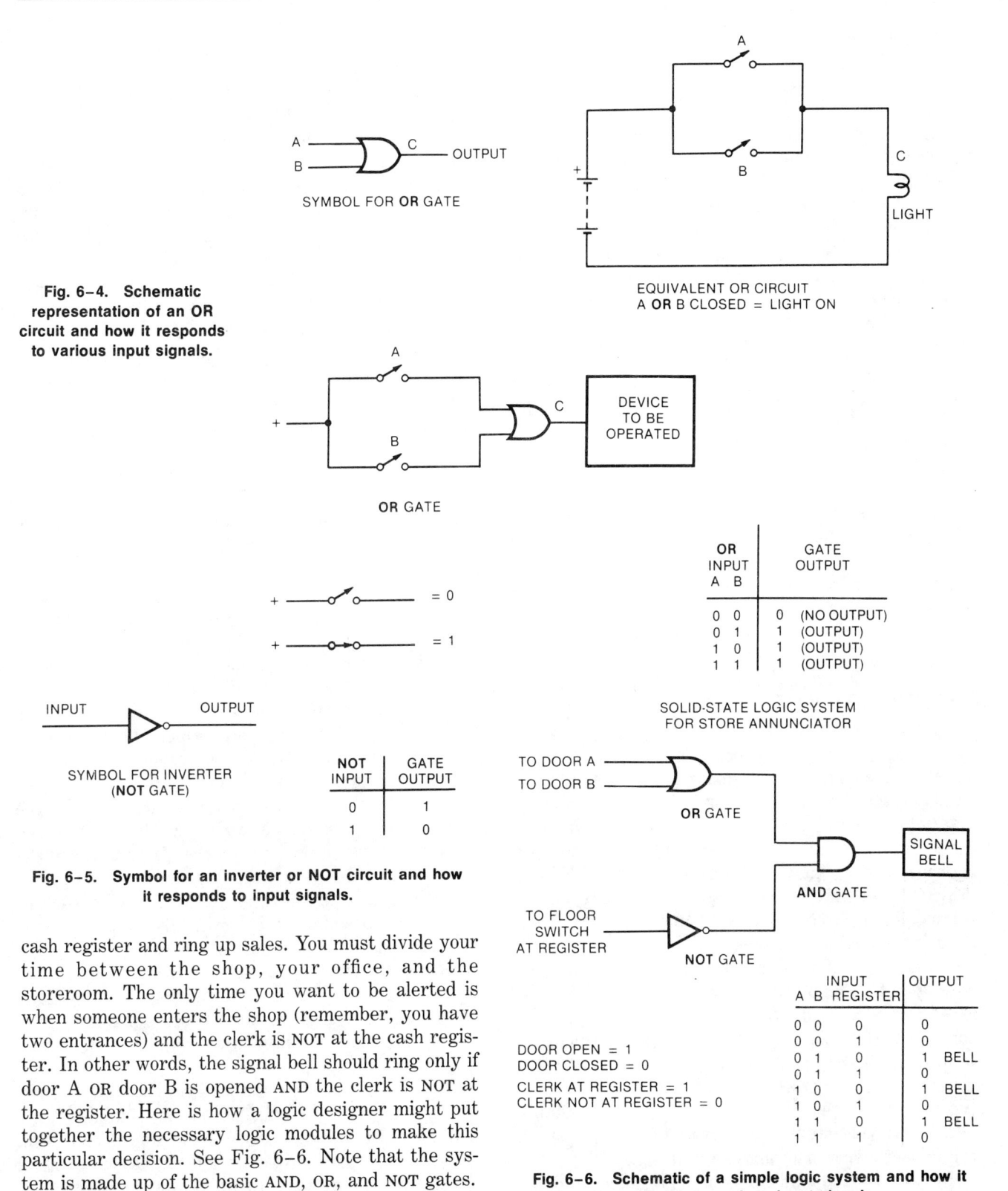

Fig. 6-6. Schematic of a simple logic system and how it responds to various input signals.

When the doors were closed, the switches would send no voltage (a logic 0 state). The floor directly in front

63

Introduction to Automotive Solid-State Electronics

of the cash register could be equipped with a mat-type switch. Such a switch would send out a voltage (a logic 1) when the clerk was at the register and, when he was away from the register, send out no voltage (a logic 0). Since we are concerned only when the clerk is not at the register, we use a NOT gate to invert the logic 0 signal to a logic 1 signal.

Let's test this logic or decision-making circuit to see if it really does what we want it to do. Assume that the clerk is by the register and that door A opens. The OR gate will now receive a logic 1 (for door A open) and a logic 0 (for door B closed). Since an OR gate will produce a logic 1 if either input is at a logic 1, we see that a logic 1 is fed to one of the inputs to the AND gate. Because the clerk is standing on the mat switch, it is producing a logic 1. However, the NOT gate inverts this to a logic 0. Therefore, the AND gate "sees" a logic 1 from the OR gate and a logic 0 from the NOT gate.

Because an AND gate cannot produce a logic 1 at its output unless *both* inputs are at a logic 1, the AND gate produces a logic 0. The signal bell does not ring, just as we had intended. For an exercise, assume other combinations of clerk positions and door openings and closings. For whatever combination you might choose, the signal bell should respond as originally intended.

Combinations of Logic Gates

The above example was a relatively simple one as far as logic systems are concerned. In real-world situations, a logic or decision-making system may employ dozens of logic gates and receive inputs from a variety of sources. The design of logic systems is a specialized branch of engineering using highly sophisticated techniques of problem solving. However, the complete system, regardless of its ultimate complexity, utilizes for the most part only the basic logic gates discussed above.

To simplify logic design, an AND gate or an OR gate is frequently combined with a NOT gate to form a NAND (for NOT AND) gate or a NOR (for NOT OR) gate. The reason for this is that it is easier, electronically, to make a NAND or NOR gate than an AND or OR gate. This is a minor point as far as we are concerned, since the resulting gates still perform the same basic functions. Just consider that the output of a NAND gate is the opposite of that of an AND gate.

To complete the family of basic logic gates, we have a specialized version of the OR gate called the "exclusive OR" gate. It is usually abbreviated EOR. As we have shown, the standard OR gate produces a logic 1 at its output if *either* or *both* its inputs are at a logic 1. The EOR works the same way, except when *both* its inputs are at a logic 1. In this case, it produces a logic 0 at its output. Although the exclusive OR function can be created using only basic NAND or NOR gates, it greatly simplifies logic circuits to use the specially designed EOR gate. In addition to decision-making systems, EOR gates are commonly used where arithmetical calculations (such as addition and subtraction) must be performed. They are well suited for such purposes.

Automotive Application of Logic Units

As a last example, let's take a look at how these basic logic units might be used in a real-world automotive application. The growing use of on-board logic systems has opened up a whole new area of decision-making circuits to better inform the vehicle's driver of impending trouble. Here is one such possibility. Although it is just a hypothetical application, it does show that with suitable input data a system can be devised to evaluate that data and come up with an intelligent decision.

When a car owner sees the red, high-temperature warning light flash on, it tends to create a near-panic situation. The driver wonders what has happened. Has the engine developed serious trouble? Has the cooling system suddenly failed? Certainly it is a situation that cannot be ignored.

However, there is one situation in which such a condition may be considered normal. When a vehicle is operating with the air conditioner on in slow-moving traffic on a hot, muggy day, it is possible for the coolant temperature to temporarily exceed the trip point of the temperature warning light. Obviously, such a situation is not a cause for panic and the remedy is simply to turn off the air conditioner. Knowledgeable drivers may be able to come to this conclusion based on their own observations. Less knowledgeable drivers may not react so calmly.

The problem could be solved with a fairly simple logic system, utilizing a few of the logic gates already discussed. But first we must provide the necessary input data to our system if we expect it to make a proper decision.

The first bit of input data we need is whether the A/C system is on or off. This can be obtained easily enough from the A/C control switch on the dash. When the A/C is on, we feed a logic 1 to our system; when it is off, we feed a logic 0 to our system.

Next, we need to know if the vehicle is moving slowly or not. This can be obtained from a speed transducer switch attached to the speedometer cable. When vehicle speed drops below say 10 miles/hour,

Microprocessors, Computers, and Logic Systems for Automobiles

the switch closes and sends a logic 1 to our control system.

We must also know if the weather is unusually hot. This bit of data can be supplied by an ambient temperature switch mounted close to the A/C condenser coils. When ambient temperature exceeds a preset level, the switch closes and sends a logic 1 to the control system.

Finally, we must know whether the coolant temperature has risen above its normal limit. This data, of course, can be supplied by the standard temperature-sensor switch mounted in the engine block. When the coolant temperature rises and the sensor trips, a logic 1 is sent to the control system.

That takes care of the input data. Now, what about the output data? In other words, what do we want our control system to tell us? To keep the system simple, let's content ourselves with two indicator lights. One will be the standard high-temperature warning light while the other will be a turn-off-the-air-conditioner light. This is the light that will come on if the overheating is caused by the use of the air conditioner. The standard warning light will come on if the overheating is caused by anything other than the air conditioner. The logic system is to be so designed that, in the event of overheating, only one or the other light will come on.

A simple way to show all the possible combinations of input conditions, and the desired output indications, is by the use of what logic designers call a *truth table*. See Fig. 6-7. From this table, it is easy to see that the only time the air-conditioner light can come on is when all inputs are at a logic 1. The only times that the overheat light can come on are when the engine temperature sensor produces a logic 1 and one or more of the other input devices produce a logic 0.

One possible logic system that can give us the desired results is shown in Fig. 6-8. In logic design there is seldom only one circuit that will give the required output. This particular circuit uses an AND gate and an EOR (exclusive OR) gate. Note that the AND gate has four inputs. However, it functions in the same manner as the previous two-input AND gates. That is, all four inputs must "see" a logic 1 before a logic 1 appears at the output.

Now let's test the system by putting 1's and 0's at the various inputs and seeing if the resulting outputs are what we want (refer to Fig. 6-7). Take the case where the engine temperature sensor trips on (logic 1), the A/C is on (logic 1), the vehicle speed is over 10 miles/hour (logic 0), and the ambient temperature is high (logic 1).

Since only three of the AND gate's four inputs are

DECISION-MAKING CIRCUIT TO DETECT ENGINE
OVERHEATING DUE TO THE A/C SYSTEM

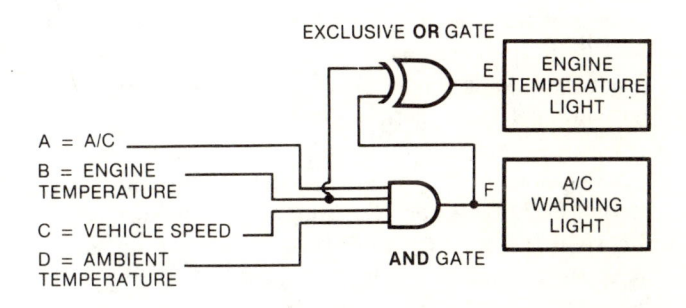

EXCLUSIVE OR GATE

A = A/C
B = ENGINE TEMPERATURE
C = VEHICLE SPEED
D = AMBIENT TEMPERATURE

E → ENGINE TEMPERATURE LIGHT
F → A/C WARNING LIGHT

AND GATE

A/C ON = 1
A/C OFF = 0

ENGINE TEMPERATURE HIGH = 1
ENGINE TEMPERATURE LOW = 0

VEHICLE SPEED HIGH = 0
VEHICLE SPEED LOW = 1

AMBIENT TEMPERATURE HIGH = 1
AMBIENT TEMPERATURE LOW = 0

Fig. 6-7. Truth table for the A/C warning-light logic system.

INPUT	OUTPUT
A B C D	E F
0 0 0 0	0 0
0 0 0 1	0 0
0 0 1 0	0 0
0 0 1 1	0 0
0 1 0 0	1 0
0 1 0 1	1 0
0 1 1 0	1 0
0 1 1 1	1 0
1 0 0 0	0 0
1 0 0 1	0 0
1 0 1 0	0 0
1 0 1 1	0 0
1 1 0 0	1 0
1 1 0 1	1 0
1 1 1 0	1 0
1 1 1 1	0 1

ENGINE TEMPERATURE LIGHT
ON WHEN E = 1
A/C LIGHT ON WHEN F = 1

Fig. 6-8. Schematic of the logic system designed to warn of engine overheating due to A/C use.

at a logic 1, its output will be logic 0. Therefore, the air-conditioning light will *not* come on. The two inputs to the EOR will see a logic 1 and a logic 0 (from the output of the AND gate). Under these conditions, it functions as a standard OR gate and produces a logic 1 at its output which turns on the engine overheat light.

Next, consider the same conditions except this time the vehicle speed is below 10 miles/hour. Now all four AND gate inputs are at a logic 1 and a logic 1 appears at the output. This turns on the air-

Introduction to Automotive Solid-State Electronics

conditioner light, telling the driver that he should temporarily turn off his air conditioning. The EOR gate sees a logic 1 at both its inputs thus producing a logic 0 at its output. This prevents the engine overheat light from coming on. (Remember, the EOR gate produces a logic 1 output only if one or the other input, but not both, is at a logic 1.)

Whatever input conditions you may select, the output result will always be as shown in the truth table of Fig. 6-7. Whether such a logic system will ever be incorporated in an actual vehicle is not important. What is important is the fact that it illustrates the basic function of logic circuits in decision-making systems. Stripped to its bare essentials, a decision-making system comprises three subsystems: (1) various input devices, (2) a logic circuit, and (3) one or more output devices. The input devices can be, and frequently are, nothing more than switches that can be either open or closed (to produce a logic 1 or 0). The output devices can be a multitude of things, from simple lights (as in our example) to solenoid-actuated valves, motors, and so forth.

In most cases, the logic circuits lack sufficient output power to operate such devices directly. Instead, they operate some intermediate device, such as a relay or a power transistor, which in turn operates the desired device. Functionally, the results are as if the logic circuits themselves were operating (or "driving") the output devices.

LOGIC SYSTEMS AND THE COMPUTER

The logic systems we have just discussed are essentially *hardware* systems. That is, they do not contain a *software* program such as is required for a computer system. They are made from discrete, "hard" parts such as AND gates, NOR gates, and the like and are designed to perform specific or "dedicated" functions. Their functions cannot readily be changed. However, for many simple automotive control requirements, such dedicated logic systems are perfectly adequate.

But when more complex logic functions are to be performed, or where it may be desirable to alter or change these functions, say from one model to the next, a true computer system is the best approach. A computer can be programmed through its software to perform the same logic functions we have just described. Also, if that software program is stored on a separate chip called a ROM (for Read Only Memory), it can be readily changed simply by unplugging it and inserting another ROM chip containing the desired program changes. Some on-board computers do just that.

As you can see, the on-board computer approach to engine control offers the car maker a rather flexible method, especially when the vehicle must respond to data from a variety of input devices or when a number of output devices must be controlled. The fact that a computer's response can be changed through software makes it especially attractive.

In a moment we will be taking a close look at just what a computer is and what makes it work. But regardless of its seeming complexity, it is still basically a logic system, although a fairly sophisticated one. To perform useful functions, it still requires (1) input data, (2) logic processing, and (3) output control. A simplified block diagram of an on-board microcomputer is shown in Fig. 6-9.

THE MICROPROCESSOR AND THE AUTOMOBILE

What is the microprocessor? In essence, it is the heart or nerve center of the microcomputer. It is the thinking part of a computer system; through it passes all the data collected from various sources. Within the microprocessor, data is acted upon, compared, manipulated, or stored for future use.

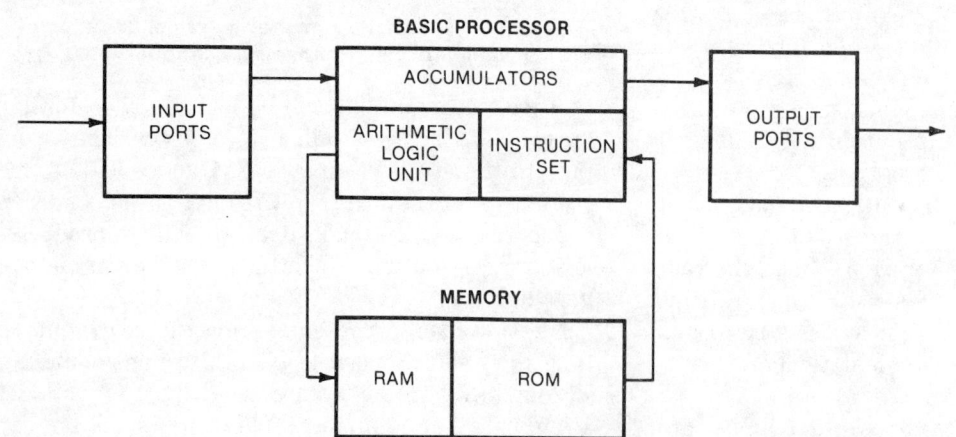

Fig. 6-9. Block diagram of a microprocessor-based on-board computer showing its essential structure.

First let us clarify a few points of terminology. A microprocessor is not necessarily a microcomputer, although with advancing technology the differences are becoming less and less apparent. Originally, a microprocessor was a major part of a microcomputer.

Today, however, microprocessors are being looked upon as microcomputers. In fact, they are being called *single-chip microcomputers*. They contain all the essential elements to make them behave as a computer, including the all important ingredient—*the program*.

All computers, from the giant "number crunchers" used in banks and large corporations to your pocket calculator, require a program. In a general purpose computer, the program can be readily changed so that different tasks can be performed. In a "dedicated" computer, such as most calculators or onboard automotive computers, the program is not readily changeable (if it can be changed at all). These computers are designed to perform one or several specific tasks, such as maintaining an engine's air–fuel ratio at a certain, predetermined level. Since *what* a computer does is determined solely by the program, let's take a closer look at this subject.

The Program

A program is what makes a computer "smart." Without a program, a computer is a pretty helpless, useless collection of parts. It can do absolutely nothing. But give it a program and it can do some wonderful and amazing things. It can turn on and off machines in a complex manufacturing operation; perform thousands of calculations in a split second; generate music; fly an airplane; tune a radio; play chess.

Give a computer the proper program and it can do just about anything you want it to do, from serious and important jobs to the trivial and the nonsensical. They can even make mistakes and do stupid things as well (as we sometimes experience or read about). However, these instances are frequently the result of poor programming, even though occasionally they can be caused by a hardware failure in the computer itself.

In talking about computers you will hear the expressions *hardware* and *software*. Hardware pertains to the actual electronic components that make up the computer circuits—the resistors, diodes, integrated circuits, and other hard parts. Software is the program that makes the hardware do what you want it to do. Software really does not exist in a tangible form unless you consider as tangible the paper on which the programmer first writes his program. The software program is simply a listing, in sequential form, of the steps or commands necessary to make a computer do the desired task. Before the computer can do anything at all, the program must be fed into it by one of several methods. We will discuss this a little later.

The program is the all important link between man and computer. A computer can never be "smarter" than the person who programs it. It can be, and is, a lot faster. And this is the true power of the computer. Although it cannot perform any calculation or operation that the programmer himself cannot perform, its processing time is measured in millionths of a second. It is this prodigious speed that has caused some to call computers, somewhat erroneously, "electronic brains." Even though they can out-calculate any human in terms of speed, they are really quite simple-minded machines.

We might say of someone who can multiply fairly large numbers in his head that "he has a mind like a computer." We may smile to ourselves when we see a small child counting on his fingers to perform simple sums. Yet a computer more closely imitates the actions of the finger-counting child than it does the man with "a mind like a computer."

Because a computer is limited to a relatively few simple operations (called *instructions*), compared to the number the human mind can perform, the program that resides within the computer must be broken down into a great number of very simple steps.

For example, we can multiply 7 by 9 in our head and get 63 in what we would call just one step. A computer cannot do this. In the first place, a computer does not recognize our symbols for the numbers 7 and 9. The computer gets around this first difficulty by having a special hardware circuit (meaning it doesn't require a software program) to translate our 7 and 9 into the *binary* system. See Table 6–1. This is the only number system the computer can understand and handle. Even though we may punch into the computer keyboard a 7 and a 9, the numbers that actually appear in the computer's processing circuit are 0111 and 1001.

This number system is necessary because the electronic circuits that will represent these numbers can be in only one of two states—either ON or OFF! If ON represents a 1, then OFF represents a 0. Each digit in a binary number is called a *bit*, and each bit requires a separate circuit, consisting of several transistors, to represent it. This is why even a small computer must contain tens of thousands of transistor circuits.

Let's see what a computer must go through just to multiply 7 by 9. First, we need a program that shows the computer how to multiply using the 24 or so instructions that have been built into the computer's

Introduction to Automotive Solid-State Electronics

Table 6-1. Binary Numbers

Decimal Number	Binary Number
0	0000
1	0001
2	0010
3	0011
4	0100
5	0101
6	0110
7	0111
8	1000
9	1001

hardware. Although they can add and subtract, most of the small computers used for automotive applications do not have an instruction that says, in effect, "multiply number *A* by number *B*." (Some of the more sophisticated computer chips do have such an instruction, but they are not widely used as yet.) However, lack of a multiply instruction does not mean that a basic computer cannot perform this or any of the other more complex functions.

Just as there is more than one way to skin the proverbial cat, so there is more than one way to multiply numbers. A simple but effective technique is by repeated addition. If we add the number 7 to itself eight times, the result will be the same as if we had multiplied 7 by 9—i.e., 63. Fortunately, there is an instruction that lets us add two binary numbers together. There is also an instruction that tests to see if a number is zero or not and another instruction that decreases a number by one each time this is requested. These, plus a few other instructions, permit us to complete the program.

It should be noted that two different programmers could conceivably come up with two different programs, since there is usually more than one way to solve a problem. This is why some programs are shorter or more efficient than others, and why some programmers are better at their jobs than others. Remember, the program must be stored bit by bit in the computer's memory if it is to tell the computer what to do. In any given computer, large or small, there is only so much memory space. An overly long or inefficient program may not fit in the allotted memory space.

Here is an example of what our program might look like. Although we have written our program in full English sentences for clarity, a programmer would do it in an abbreviated shorthand style following strict programming rules.

STEP	OPERATION
1	Take the data (the number 7 in binary form) and place in Accumulator A. (Note: An accumulator is a special storage space within the computer's hardware where data can be acted upon. This is where much of the "computing" takes place.)
2	Take the next piece of data (the number 9) and place in Accumulator B.
3	Decrease the number in Accumulator B by 1.
4	Take the number in Accumulator A and store it in Memory Location 100.
5	Add the number in Memory Location 100 to the number in Accumulator A.
6	Decrease the number in Accumulator B by 1.
7	If the number in Accumulator B is more than zero, go back to step 5 and repeat the program. If it is zero, go to the next step.
8	Display the number in Accumulator A. (Note: The binary number in Accumulator A must first be converted back to a conventional number so that we can easily recognize it. This is done by the computer's hardware.)
9	The job is done. Stop.

A computer doesn't even know when it's through computing. It has to be told to stop. If everything went the way we had planned, and there were no "bugs" in our program, the computer would flash back to us the number 63, the answer to the question: How much is 7 times 9. If this seems like a cumbersome and roundabout way to do a relatively simple job, it is. But then we never said computers were smart, just fast—very, very fast. In an average-speed computer the above program would take approximately 100 microseconds. A microsecond is one millionth of a second.

Data Storage

In addition to performing arithmetic functions, such as might be required by a "trip computer" (see Chapter 8) for miles per gallon, a computer can also store data, look up data in a table, and perform all the logic functions discussed previously. In the little multiplication program we just finished, note that we instructed the computer to store some data in Memory Location 100. The number 100 is the *address* of a specific location in what is called a Random Access Memory (RAM). It is like a street address that identifies a particular house. Typically, each address can store eight bits of data or, as computer programmers say, a *byte* of information (8 bits = 1 byte). A small, on-board computer can store perhaps up to 128 bytes of information. The RAM is a temporary memory used primarily to store data from the various input devices prior to being acted upon by the program. It is also used to store output data that is to be sent to some output device. Whatever data is stored in RAM is lost when the power is removed, usually by turning off the ignition key. The "trouble codes" discussed in Chapter 7 are stored in RAM. See Fig. 6–9.

Computers have another type of memory called a Read Only Memory (ROM) that is permanent. That is, its memory is not lost when the power goes off. This is where the all-important program is stored. It may also contain some *look-up tables* which are useful because they can often save computation time and program steps. For example, a computer that controls the amount of distributor advance can have this information stored in a table, a different amount of advance at each address. The input data that determines distributor advance (e.g., engine speed, manifold vacuum, temperature) is so coded that it produces the correct address for the necessary amount of advance. In other words, instead of the computer "computing" the required advance, it simply looks it up in a precomputed table. However, not all engine-control functions can be handled in this manner. Some must be computed.

How do the program, look-up tables, and other permanent data get put into a ROM? There are several methods. When the ROM is manufactured, each address is "masked off" to produce the necessary bits (the 0's and 1's) that make up the program or the look-up table. This method is economically feasible only in large quantities. Once programmed, the ROM cannot be changed. If the ROM is made on the same chip that contains the microprocessor (i.e., a one-chip microcomputer), the whole computer must be altered if a program change is needed. This is why ROMs are sometimes put on separate chips.

Another type of ROM is the Programmable Read Only Memory (PROM). These are manufactured with a blank memory. The necessary program or table is then "burned in" with a special programming machine. Of course, once a PROM has been programmed, it cannot be changed. But the advantage of a PROM is that it can economically be produced in small quantities. Thus program changes for various vehicles can be made readily.

There is yet another type of ROM called an EPROM (Erasable PROM) in which the program can be erased by exposure to ultraviolet light. Such ROMs can be programmed many times. They are used primarily for development work or where it is felt the program may have to be changed.

On-Board Computers

The primary function of on-board computers is to achieve the necessary emission control while at the same time maintaining the maximum fuel economy, consistent with emission and drivability requirements. We have mentioned the role of the computer in controlling distributor advance. They are also used to control the EGR valve discussed in Chapter 5. Another important function, that of feedback carburetion, will be discussed next.

It should be remembered that all on-board computers do not perform all the functions we have described. Some perform only one function, such as spark control. Others handle practically all engine functions. As computers become more sophisticated and as memory-storage capabilities increase (meaning longer and more complex programs), more engine and even nonengine tasks will be assigned to them.

FEEDBACK CARBURETORS

In a conventional carburetor, the delivered air–fuel ratio is determined primarily by throttle position and intake manifold vacuum. The basic assumption is that as the throttle opens, or the manifold vaccum drops, the engine is under a greater load and therefore requires a richer mixture. The carburetor's metering rods, or power enrichment valve, move in such a manner as to allow more fuel to enter the discharge nozzle. No attempt is made to monitor the actual air–fuel ratio being delivered to the engine.

On the other hand, a feedback carburetor does monitor the air–fuel ratio and corrects the mixture to maintain essentially a fixed ratio. It accomplishes this through a closed-loop system by sampling the exhaust gas (which reflects the true air–fuel ratio) and then sending a correction signal back to the carburetor. Hence the name *feedback carburetor*. Several systems are employed by the various manufacturers, although the end results are the same—maintaining the mixture ratio at the stoichiometric point.

The major component in the feedback system is the *oxygen sensor*. See Fig. 6–10. The sensor is placed in

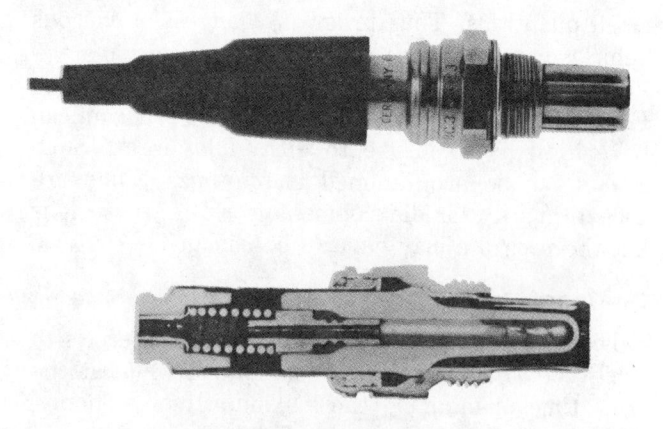

Introduction to Automotive Solid-State Electronics

Fig. 6-10. Typical oxygen sensor.
Courtesy Busch

the exhaust system, usually in the exhaust manifold, where it is exposed directly to the exhaust gases. It is designed to respond to the oxygen content in the exhaust-gas sample by producing a voltage signal. However, before we discuss how the feedback system works, we should explore just what it is that the system is designed to accomplish.

Purpose of the Feedback Carburetor

The purpose of the feedback carburetor system is to maintain the air–fuel ratio at the stoichiometric point. The stoichiometric ratio is the mixture that results in theoretically perfect combustion (see Chapter 5). The actual ratio varies with the particular fuel being used but is typically between 14:1 and 15:1 (i.e., 14 parts air by weight to 1 part fuel by weight). When stoichiometric mixtures are burned, they produce the least amount of unburned hydrocarbons (HC) and carbon monoxide (CO). In theory, the byproducts of combustion from such mixtures should consist only of carbon dioxide (CO_2) and water vapor (H_2O) In practice, small amounts of the unwanted HC and CO do appear. These trace amounts, of course, are easily handled by the emission-control systems, principally the catalytic converter.

Oxygen Sensor

A characteristic of mixtures near the stoichiometric point is the amount of oxygen produced during the combustion process. If the mixture is slightly on the rich side of the stoichiometric point, very little oxygen appears in the exhaust-gas sample. However, if the mixture is slightly leaner than stoichiometric, the oxygen content increases substantially. The oxygen sensor is the device that detects the presence of oxygen in the exhaust and determines whether the mixture is richer or leaner than the stoichiometric level.

The sensor consists of a hollow ceramic body made from zirconium dioxide with platinum electrodes on both the inside and outside. See Fig. 6-11. The inside electrode is exposed to the air while the outside electrode is exposed to the exhaust gases. When the outside electrode is surrounded by exhaust gas largely free of oxygen (i.e., from a mixture richer than stoichiometric), a voltage potential of up to 800 millivolts (1000 millivolts = 1 volt) develops between the two sensor electrodes. But when the exhaust gas contains a large amount of oxygen (i.e., from a mixture leaner than stoichiometric), the electrode potential drops to a very low level. Thus the oxygen sensor provides essentially a digital output (either a voltage signal or no voltage signal), depending on whether the mixture is above or below the stoichiometric point.

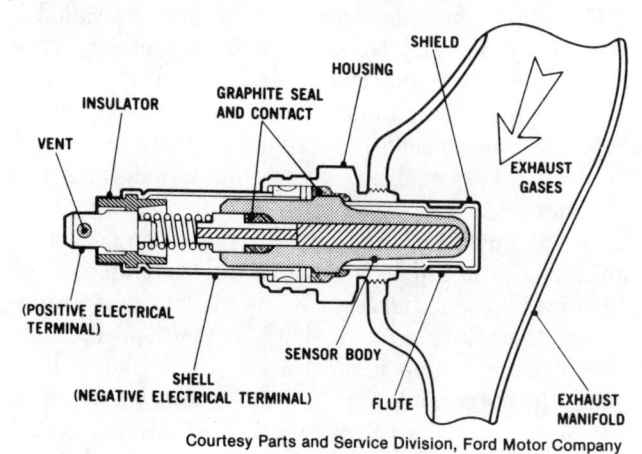

Courtesy Parts and Service Division, Ford Motor Company

Fig. 6-11. Cross section of oxygen sensor showing its main components.

This signal is sent back to the computer which sees it as either a logic 1 or a logic 0. The computer processes the data according to its stored program and then outputs a control signal to a device attached to the carburetor that can vary the air–fuel ratio. It is easy to see why it is referred to as a feedback system.

The device that varies the air–fuel ratio can be either a metering rod that moves in or out of a jet or an air bleed control that can be quickly adjusted to alter the amount of air being bled into the high-speed discharge nozzle. Although the methods differ, the results are the same.

A system used by General Motors employs a mixture-control solenoid to move a metering rod in and out of a jet. See Fig. 6-12. Note that the control solenoid also controls the amount of air being bled into the system. The main metering system actually contains two calibrated jets, the lean jet which is always in the system and the rich jet through which

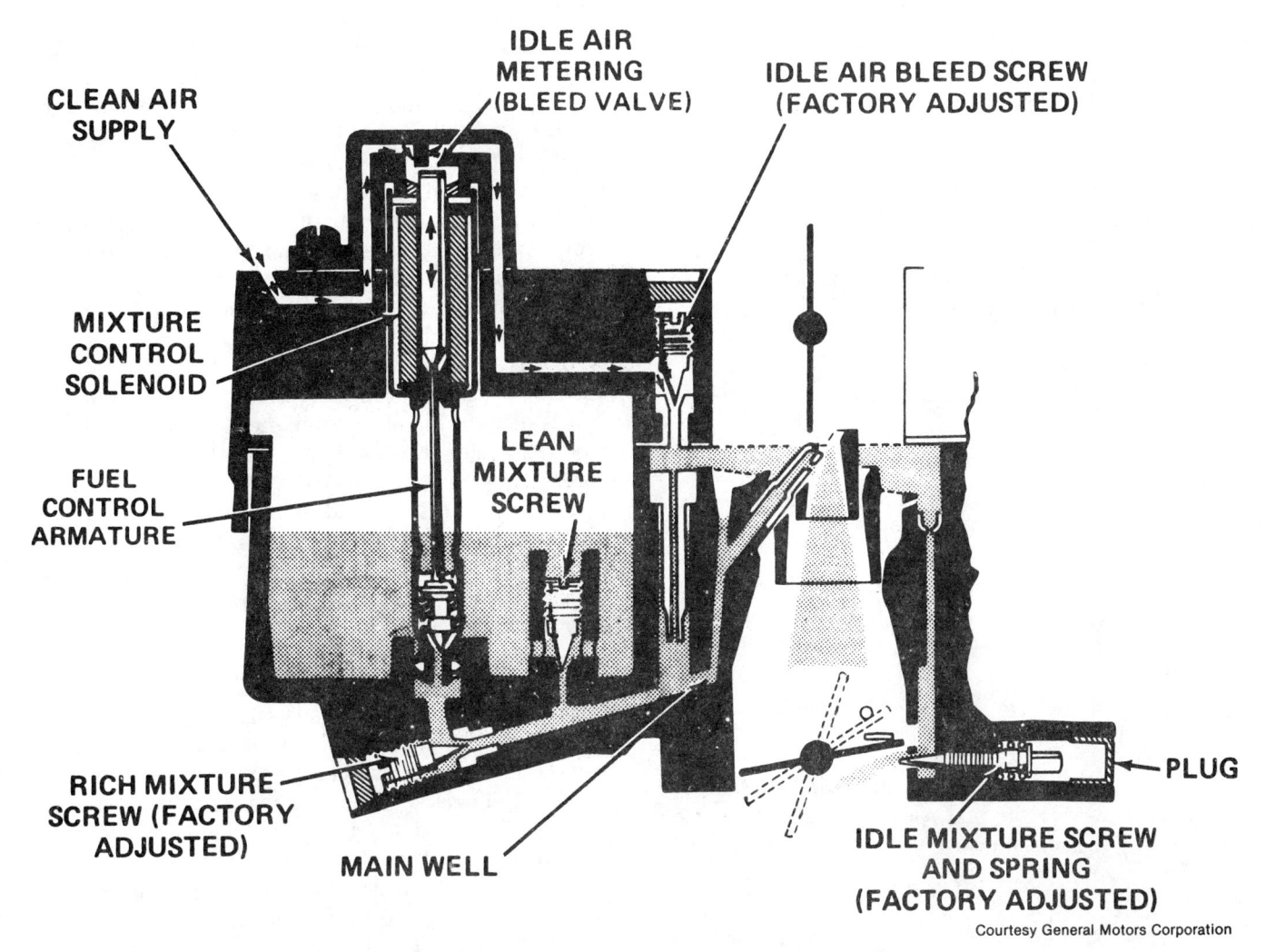

Microprocessors, Computers, and Logic Systems for Automobiles

Fig. 6-12. Solenoid control valve used in a feedback carburetor.

fuel passes only when the mixture-control solenoid allows it. The current to operate the solenoid is controlled by the signal from the computer, which in turn receives its input from the oxygen sensor.

The Ford system uses a stepping motor that can be rotated to a large number of positions. A stepper motor, unlike an ordinary motor, does not turn continuously but, rather, is set to specific positions depending on the signal fed into it: it can never make more than one revolution. Attached to the shaft of the stepper motor is a threaded member that, when rotated, moves a metering valve in or out to allow more or less air to pass through. See Fig. 6-13.

ELECTRONIC SPARK TIMING

In addition to carburetion control, the on-board computer is usually called upon to control the spark timing. In some systems, this is the sole function of the computer. Spark timing plays a vital role, not only in engine performance, but in emission control and fuel economy.

In the past, spark timing, for speeds above idle, had been controlled by two mechanical subsystems: (1) the mechanical advance unit and (2) the vacuum advance unit. The mechanical advance took care of the spark advance solely on the basis of engine speed: the faster the engine, the greater the mechanical advance. The vacuum advance took care of the spark timing based on engine load. It did this by sensing intake manifold vacuum, which is a sensitive indicator of engine load.

However, when it is necessary to achieve the maximum fuel economy consistent with required emission control and good drivability, additional factors must be considered. For example, altitude, barometric pressure, and engine temperature should be taken into account. These additional parameters

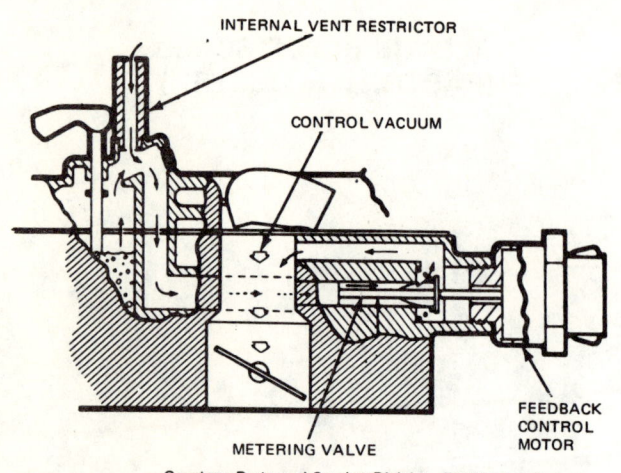

Introduction to Automotive Solid-State Electronics

Fig. 6-13. Stepper motor used to control the mixture in a feedback carburetor.
Courtesy Parts and Service Division, Ford Motor Company

were beyond the capabilities of the basic vacuum and mechanical advance mechanisms. Only a computer-based system could evaluate all the varied input data and arrive at the optimum spark advance for any given set of conditions.

A computer-based system dedicated to spark-timing control is shown in Fig. 6-14. In this system, the distributor itself is devoid of the familiar vacuum and mechanical advance mechanisms. All that remains is the reluctor and pickup coil assembly. The spark advance will be controlled by a signal from the on-board computer.

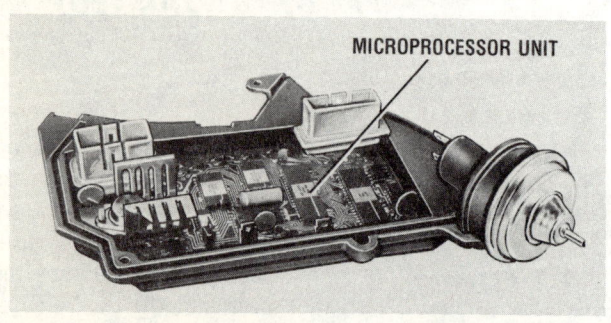

Fig. 6-14. Cutaway view of microprocessor-based on-board computer for electronically controlling spark timing.
Courtesy Chrysler Corporation

In order for such a system to function properly, it must receive input data from a variety of sources. First, it must know when a particular piston is at top dead center. It receives this data from the reluctor–pickup coil assembly in the distributor. Second, it must know the speed of the engine. This data is also received from the distributor; the computer measures the time interval between each "tooth" on the reluctor. In addition, the computer must know engine temperature (actually, whether the temperature is above or below a certain level) and the position of the throttle plates. All of this data is fed into the computer by the various input sensors. An overall schematic of this system is shown in Fig. 6-15. Note that the vacuum transducer connects to the intake manifold.

It is now the job of the computer to evaluate all the input data and output a signal to the coil to produce a spark at just the right moment. Whether the computer looks up the required spark advance data in a look-up table or actually computes it based on a formula contained in its program is not important. The important point is that it can arrive at the correct advance faster and more precisely than the earlier mechanical systems could.

Another spark-control system, called Electronic Spark Timing (EST), is shown schematically in Fig. 6-16. Note the barometric pressure sensor. The electronic module that controls the ignition coil is shown in Fig. 6-17. Note the additional terminals on this module compared to the four terminals found on noncomputer-controlled systems.

As mentioned earlier, an on-board computer may control one or several engine functions, depending on the emission and fuel requirements of the particular engine. The important thing to remember about on-board computer systems is that they are essentially processors of data. They collect information from various input sensors, process it, and then output control signals to the various engine-control devices (e.g., EGR valves, carburetor, and ignition coil). When trouble develops, rather than suspect the somewhat mysterious microprocessor unit, be sure the input devices are sending the correct data and that the output devices are capable of responding to the signals from the computer. If you take this approach to troubleshooting, your job should be much easier and less frustrating.

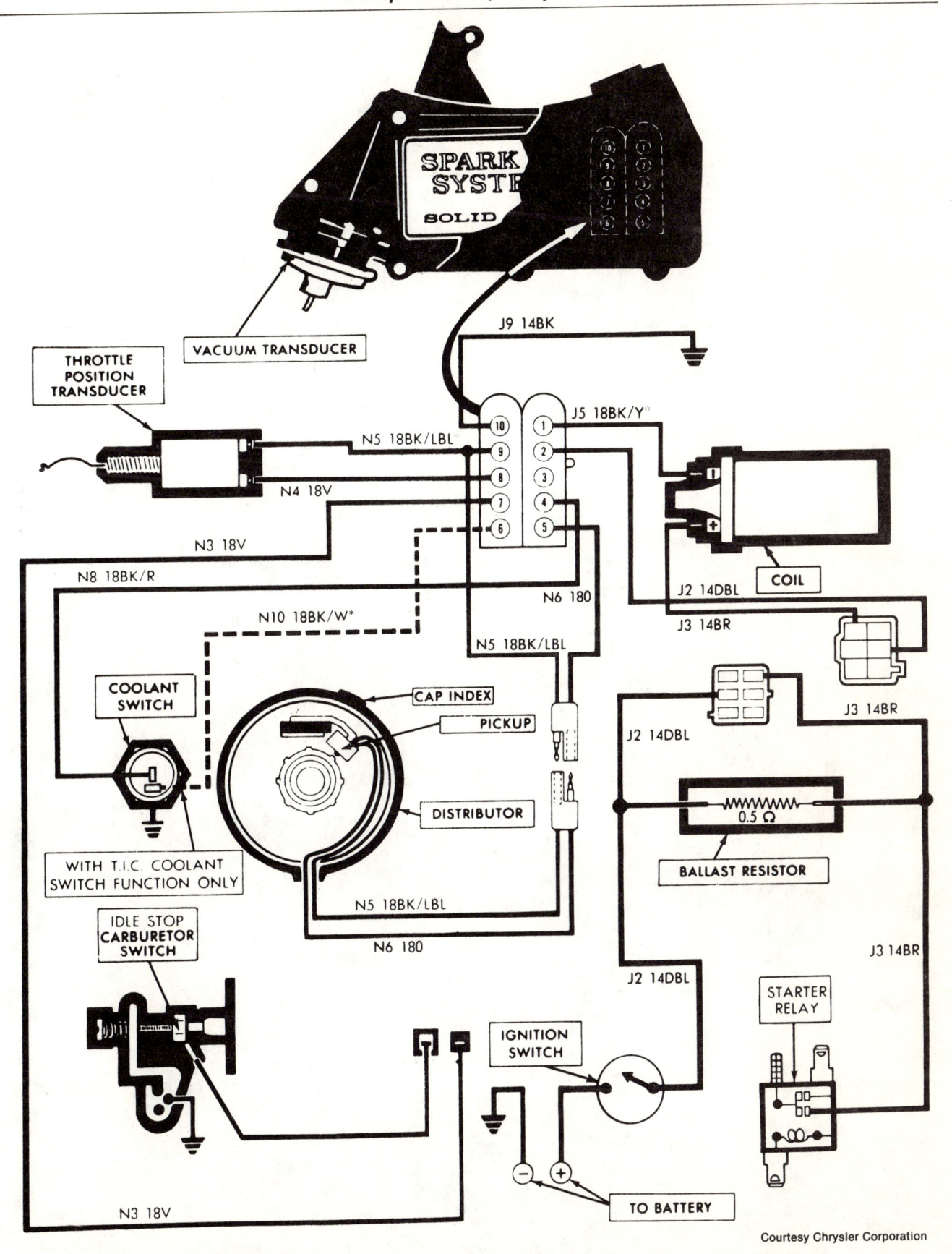

Microprocessors, Computers, and Logic Systems for Automobiles

Fig. 6-15. Schematic showing the input devices connected to the computer illustrated in Fig. 6-14.

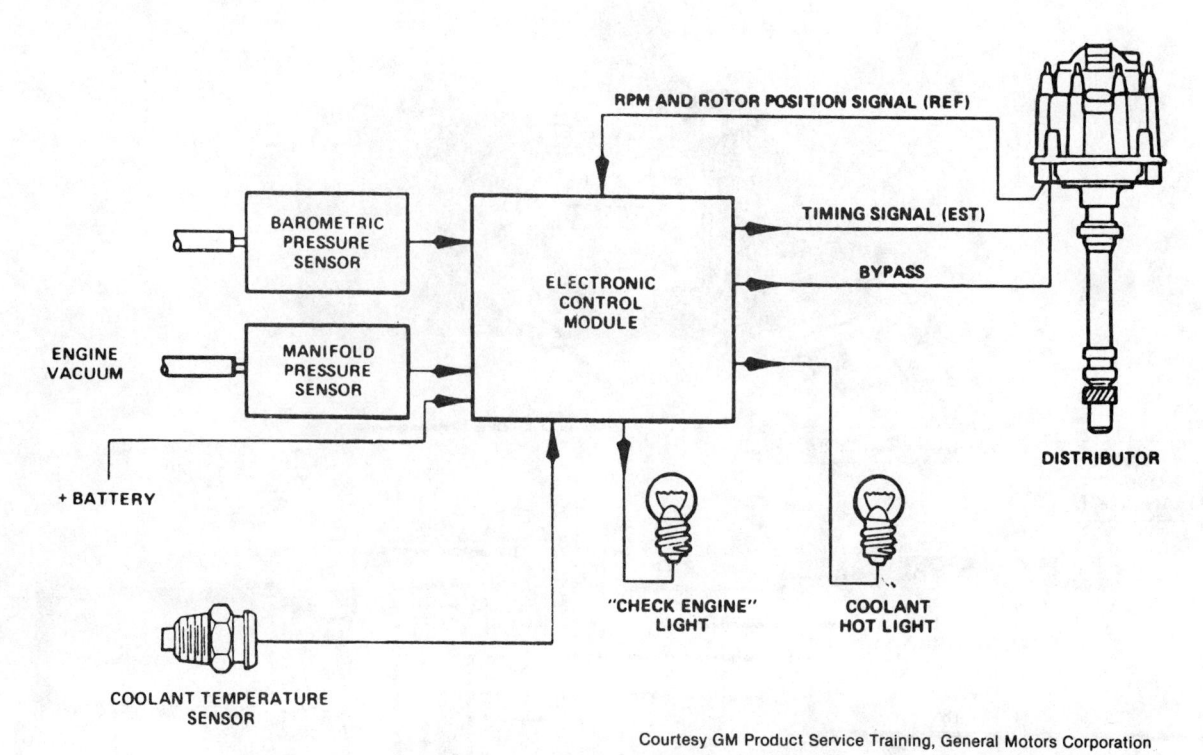

Introduction to Automotive Solid-State Electronics

Fig. 6-16. Block diagram of an Engine Spark Timing (EST) control system.

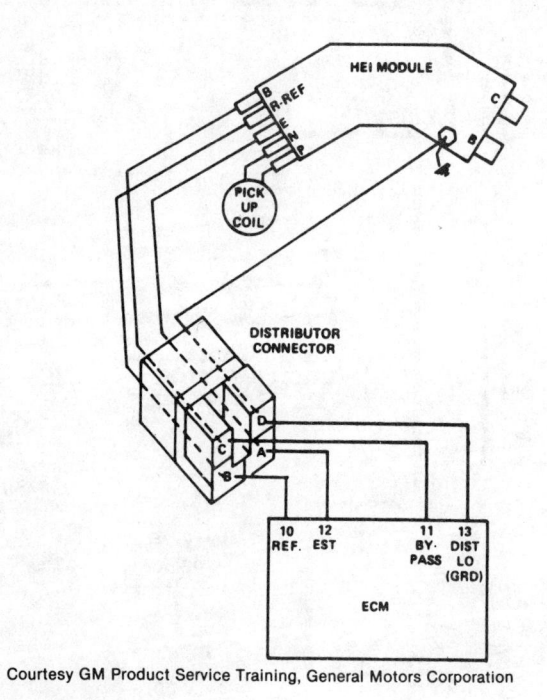

Courtesy GM Product Service Training, General Motors Corporation

Fig. 6-17. Ignition-control module used with the EST system of Fig. 6-16.

Courtesy GM Product Service Training, General Motors Corporation

CHAPTER 7

On-Board Diagnostic Systems

As we have seen in the previous chapters, many of the engine-control functions are now being performed by computer (microprocessor) based systems. These are the same functions that were formerly performed by mechanical, electromechanical, or vacuum-mechanical devices. For example, the vacuum advance function (which set the amount of timing advance based on engine load) was accomplished by a vacuum-actuated diaphragm connected to the distributor breaker plate. The vacuum used for this purpose was taken from the intake manifold, since this is a sensitive indicator of engine load. The entire advance function, however, was mechanical.

Another example would be the enrichment function in the carburetor. High speeds and heavy loads demand a richer mixture, a function that may be accomplished either by increasing the effective size of the fuel metering jet or by opening up an auxiliary metering jet in addition to the main jet. Again, this function was achieved by purely mechanical means, such as tapered metering rods inserted into the main jet and mechanically connected to the throttle valve. As the throttle opened, the metering rods would rise and allow more fuel to flow past the main jet. In this case, the design assumption was that the needed enrichment was directly proportional to throttle position. While this was generally true, it did not allow any "fine-tuning" of operating conditions.

The point of these examples is that the once all-mechanical functions are being duplicated and enhanced by electronic means in today's vehicles. Although it is true that the actual device used to accomplish the desired function (e. g., carburetor mixture enrichment) is still mechanical, it is being driven or controlled electronically. In other words, it is being controlled by an electronic decision-making system.

DIAGNOSTIC CAPABILITY

Troubleshooting mechanical systems is a lot different from troubleshooting electronic systems. With mechanical systems, we can often see what is or is not taking place. With electronic systems, visual diagnostics is seldom helpful. If it is a simple electronic system, we can usually localize the problem with a few electrical measurements. But if it is a complex system, such as a computer-based system, it will usually require a much different approach. Fortunately, these systems will usually have what is known as on-board diagnostic capability. This means that in addition to the basic engine-control program built into the system, a diagnostic program is also incorporated. In essence, the diagnostic program built into the system is designed to test itself for certain anticipated failures.

It does not mean, at the present state of the art, that any and all problems can be detected, but rather that specific groups of likely problems can be spotted or localized. What it does mean is that the system can give the troubleshooter valuable data as to the nature and possible location of the problem. It will still be necessary for the technician to apply more or less conventional troubleshooting techniques to actually pinpoint the exact cause of the problem. It should be remembered that the diagnostic program incorporated within the system can perform no more tests than the system designer chose to include. As a result, some on-board diagnostic programs will be more detailed or cover more contingencies than the programs included by others.

Self-Testing Features

To see how such an on-board diagnostic system might work, let's consider the self-testing features of General Motors' C-4 System (Computer Controlled Catalytic Converter). As we go through this system, keep in mind the basic configuration of a computer-based engine-control system. It receives input data from various sensors, evaluates it, and then sends a control signal to an output device. In this case, the output device is the mixture control solenoid for the carburetor. The system can also operate in either of two modes: (1) closed loop or (2) open loop. These will

Introduction to Automotive Solid-State Electronics

be discussed in more detail after we describe the design of the overall system.

Fig. 7-1 shows a block diagram of the major parts of the C-4 System. The Electronic Control Module (ECM) is the decision-making portion that receives input data from the various sensors, in particular the oxygen sensor. (See Chapter 6). Depending on the signal being sent by the oxygen sensor, the ECM will output a variable strength signal to the mixture control solenoid on the carburetor, which in turn varies the air–fuel ratio. The strength of the signal to the mixture control solenoid is obtained in a digital manner by varying the relative "on time" of the signal pulse to its "off time." In other words, the strength of the signal pulse is determined by controlling its *on* or *dwell* period. So similar is this to the dwell action of breaker-point ignition systems, that a standard dwell meter is used to diagnose problems in the mixture control circuit. We will cover this aspect of troubleshooting a little later.

Closed-Loop System—From the description we have just given and the diagram shown in Fig. 7-2, you will note a definite *loop* pattern. The flow of input–output data continues to circulate within this loop such that the results of the mixture control solenoid will be "seen" by the oxygen sensor, which, in turn, will direct the action of the mixture control solenoid. This sequence of events is called a *closed-loop* system. Such a system is self-adjusting; it keeps sampling the results of its actions and making corrections until the desired level of control is achieved. This mode of system operation is also said to have feedback.

When the operating conditions of the engine change, say a change in speed, the closed-loop operation of the system will again seek to correct the mixture ratio to the desired (stoichiometric) level. Because of the speed at which the computer operates, the correction process can take place almost instantaneously.

The major factor limiting the mixture correction time is the time it takes a given mixture to move from the carburetor, through the four cycles of the engine and the exhaust system to the location of the oxygen sensor.

Open-Loop System—The opposite of closed-loop operation is *open-loop* operation. In essence, the system is operating without feedback information from the oxygen sensor. This is necessary and desirable under certain engine modes such as wide-open throttle or cold-engine operation. Under these conditions, the mixture control solenoid receives a fixed signal designed to produce the level of enrichment (richer than stoichiometric) that is needed for proper engine operation. When the engine returns to more-normal operating modes, the system then reverts to more-normal closed-loop operation. The computer is informed of these special engine requirements through

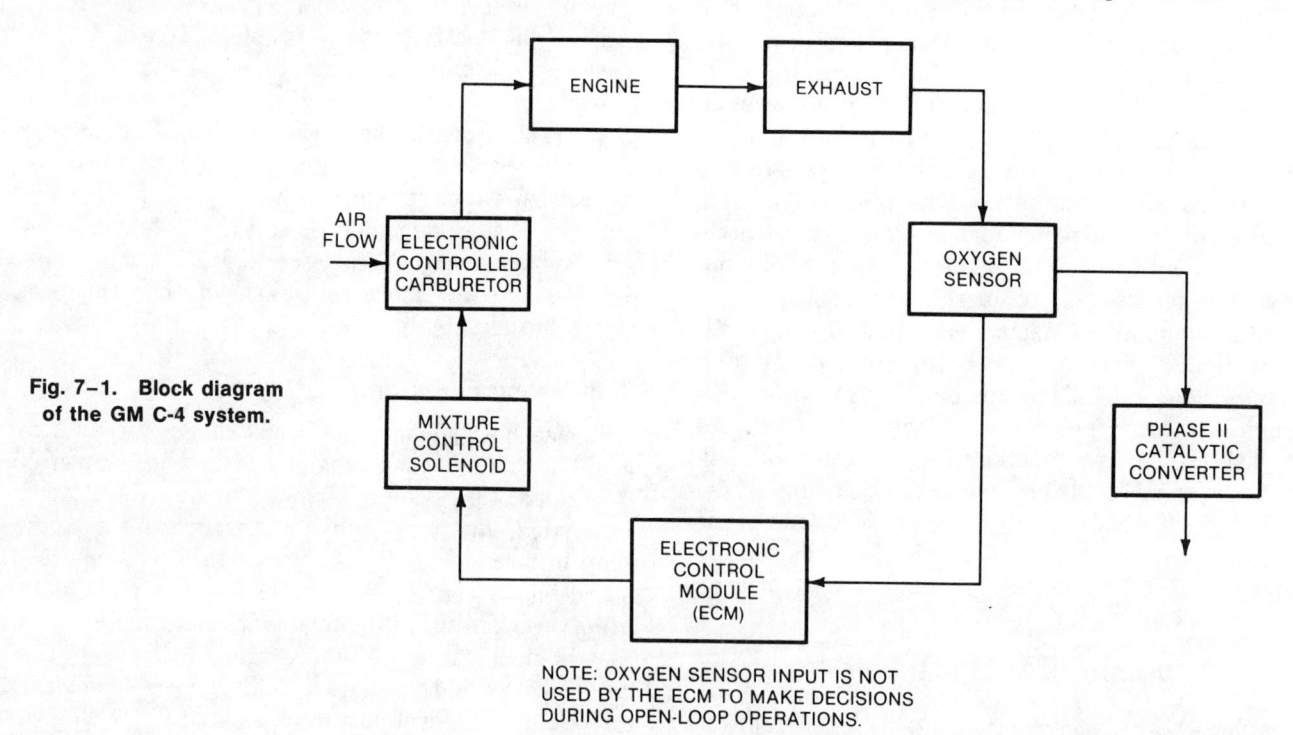

Fig. 7-1. Block diagram of the GM C-4 system.

NOTE: OXYGEN SENSOR INPUT IS NOT USED BY THE ECM TO MAKE DECISIONS DURING OPEN-LOOP OPERATIONS.

Courtesy GM Product Service Training, General Motors Corporation

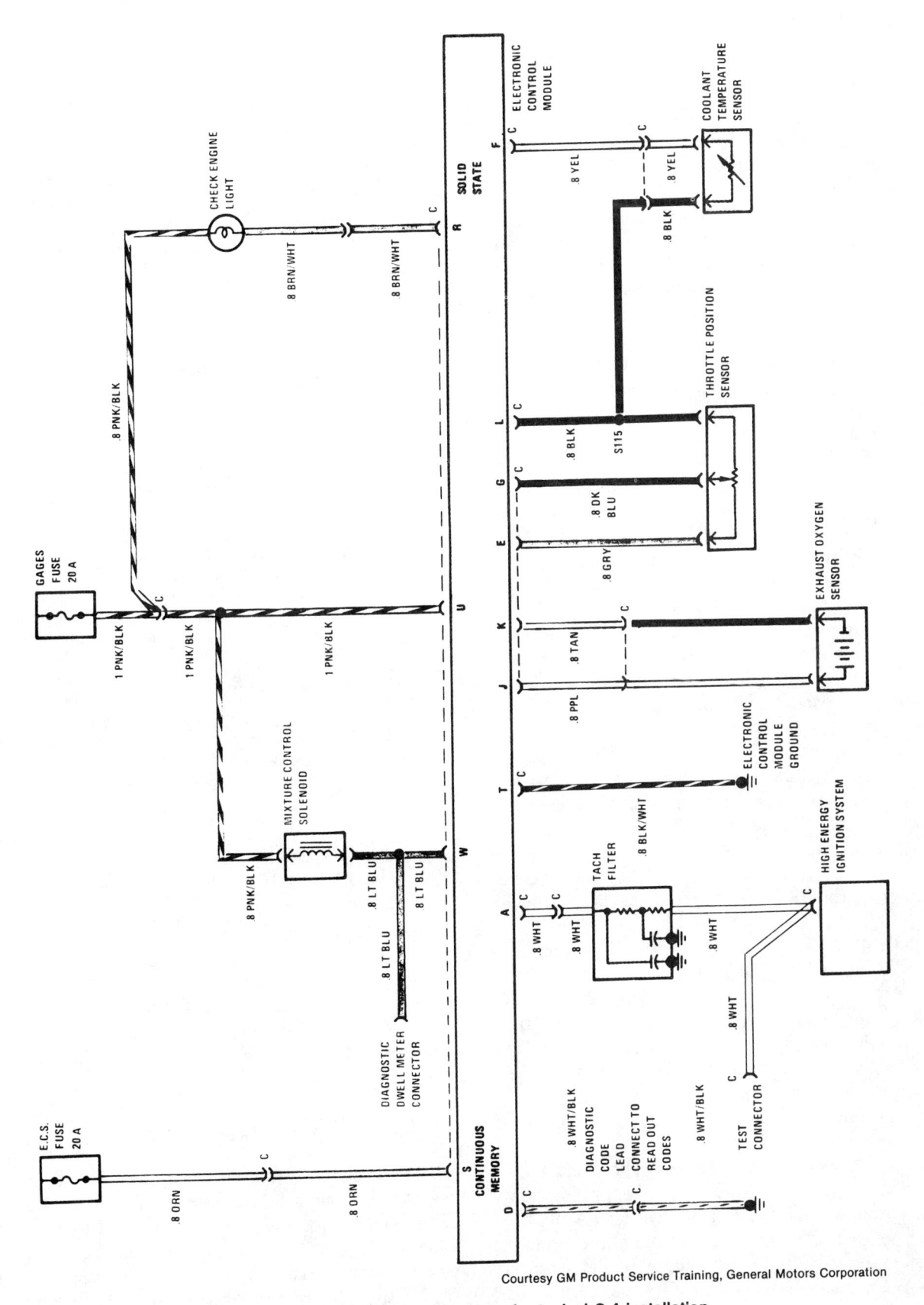

Fig. 7-2. Schematic diagram of a typical C-4 installation.

additional input devices such as throttle position and engine temperature sensors.

Self-Testing Capabilities

Now that we have seen how the basic C-4 system operates, let's consider its built-in diagnostic or self-testing capabilities. The heart of this subsystem is a *check engine* light on the dashboard which is designed to come on whenever certain faults occur in the overall system. This informs the driver that something is wrong (although it does not tell the driver what) and that the vehicle needs service. This light will normally come on (as a bulb check) when the ignition switch is first turned on and prior to starting the engine. This is the same action that occurs with standard warning lights. The light is normally off, unless a fault develops, during engine operation.

When a fault does occur, the same light can be used to localize the problem area. This is done by grounding a special diagnostic test lead located under the dash and observing the action of the light. Depending on the type of trouble, the light will flash a two-digit code number. (See Table 7–1.) For example, if the light flashes a "14" (one flash followed by a pause and then four more flashes), it indicates a problem in the coolant sensor circuit. The service technician can then refer to a troubleshooting chart for a code 14 problem and proceed to localize the problem with more or less standard troubleshooting techniques.

Diagnostic Subsystem—There is always the question: Is the diagnostic subsystem itself working properly? There is nothing special in the diagnostic portion of the system to make it foolproof. It too can fail. To assure us that this has not happened, the diagnostic subsystem can also check its own functioning. If this proves to be good, we know that any subsequent trouble code that may be flashed will be the true code.

To check the diagnostic portion itself, turn on the ignition but do not run the engine. Ground the special test lead and observe the check engine light. If it flashes a code 12 (one flash followed by a pause and then two additional flashes), you will know that the diagnostic program is working. After a longer pause, the code number will repeat itself two more times. This cycle will continue until either the engine is started or the ignition is turned off.

Codes—Other than this check, the fault code number must be obtained while the engine is running. In other words, ground the test lead while the engine is operating and note the action of the light. It is possible for more than one fault to be detected. If this happens, the first fault code number will be flashed three times, followed by the next code number. This sequence will then repeat itself.

Table 7–1 is a list of the trouble codes and the circuits to which they pertain. It should be noted that there are two code 12 indications. The first, as described above, is to ascertain whether the diagnostic program in the computer is functional and is obtained when the test lead is grounded with the ignition on but the engine *not* running. The second, which indicates a particular fault, is obtained with the engine running.

Table 7–1. C-4 System Trouble Codes

Code	Problem
12	No tachometer signal to the ECM.
13	Oxygen sensor circuit. The engine has to operate for about 5 minutes (18 minutes on 3.8L V6) at part throttle before this code will show.
14	Shorted coolant sensor circuit. The engine has to run 2 minutes before this code will show.
15	Open coolant sensor circuit. The engine has to operate for about 5 minutes (18 minutes on 3.8L V6) at part throttle before this code will show.
21	Throttle position sensor circuit on 2.8L V6.
21 and 22	(At same time) Grounded WOT (Wide-Open Throttle) switch circuit. 3.8L V6 and 2.5L L4.
22	Grounded closed throttle or WOT switch circuit. 3.8L V6 and 2.5L L4.
23	Open or grounded mixture control solenoid circuit.
51	On service unit, check calibration unit installation.
52 and 53	Replace ECM (Electronic Control Module).
54	Faulty M/C (Mixture Control) solenoid and/or ECM.
55	Replace ECM on 3.8L V6 and 2.5L L4.

Once the trouble code number has been obtained, the service technician can refer to the appropriate troubleshooting chart for further tests to pinpoint the problem. A typical troubleshooting chart is shown in Fig. 7–3.

TESTING EQUIPMENT

The use of the troubleshooting charts requires the use of only standard-type test instruments: ohmmeter, dwell meter, tachometer, voltmeter, vacuum gauge, and jumper wires. As mentioned earlier, the dwell meter is not used for ignition purposes but rather to measure the "duty cycle" of the mixture

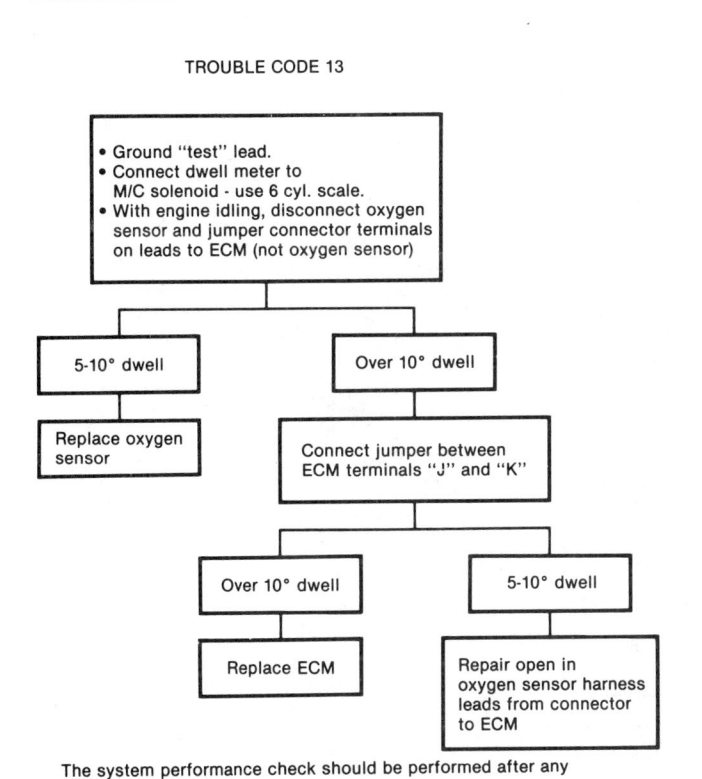

On-Board Diagnostic Systems

TROUBLE CODE 13

- Ground "test" lead.
- Connect dwell meter to M/C solenoid - use 6 cyl. scale.
- With engine idling, disconnect oxygen sensor and jumper connector terminals on leads to ECM (not oxygen sensor)

```
        ┌──────────────┬──────────────┐
   5-10° dwell              Over 10° dwell
        │                         │
 Replace oxygen          Connect jumper between
    sensor               ECM terminals "J" and "K"
                                  │
                    ┌─────────────┴─────────────┐
               Over 10° dwell              5-10° dwell
                    │                            │
               Replace ECM              Repair open in
                                        oxygen sensor harness
                                        leads from connector
                                        to ECM
```

The system performance check should be performed after any repairs to this system have been made.

Courtesy GM Product Service Training, General Motors Corporation

Fig. 7-3. Typical troubleshooting chart.

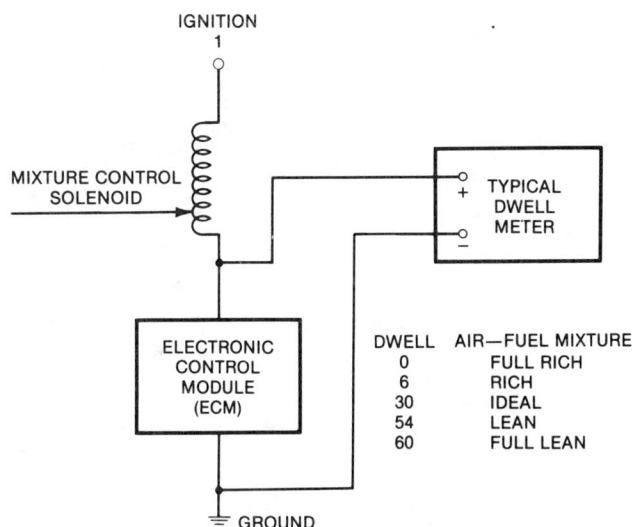

control solenoid. The standard hookup is shown in Fig. 7-4. Note that the six-cylinder dwell scale or switch position is used, regardless of the number of cylinders in the engine. This will produce a reading between 9 and 60 (ignore the degree signs) which will indicate the amount of leaness or richness of the air-fuel mixture. Depending on the specific test being made, as indicated in the troubleshooting charts, the meter should read within a specified range. Most dwell meters will work in this application. However, if one causes a change in engine operation when connected, it should not be used.

SYSTEM PERFORMANCE CHECK

Since it would be impossible to include in the diagnostic program every possible type of system failure, the absence of a trouble code does not mean that a problem does not exist. When a complaint is suspected, but no trouble code can be obtained, it will be necessary to make a System Performance Check. See Fig. 7-5. A Diagnostic Circuit Check procedure is also shown in Fig. 7-6. As discussed earlier, a procedure of this type can be used to verify the functioning of the diagnostic program or to localize any problems in this area.

DWELL	AIR–FUEL MIXTURE
0	FULL RICH
6	RICH
30	IDEAL
54	LEAN
60	FULL LEAN

Courtesy GM Product Service Training, General Motors Corporation

Fig. 7-4. Dwell meter check of mixture control solenoid.

One of the biggest problems any troubleshooter has to contend with is that of the intermittent problem. If the problem does not exist at the time the system is being checked, it probably will not be detected. In the C-4 system such a situation will manifest itself in the following manner: The driver notices that the check engine light comes on (indicating that a fault has occurred) and then goes off (indicating that the fault is intermittent and not occurring at that particular moment). If he should drive immediately into a service facility, a service technician could determine the trouble code number even though the check engine light is out. He could do this *provided* the engine has *not* been turned off following the fault indication.

This is possible because the trouble code number is stored in the diagnostic program memory (even though the check engine light is off) for as long as power is supplied to the ECM (Electronic Control Module). Turning off the ignition removes the power from the ECM and whatever was in its temporary memory storage is lost. (The ECM has two types of memory: one for temporary storage of input data that changes from moment to moment and another for the permanent storage of the actual program that controls the ECM.)

Importance of Check Engine Light

However, even though the trouble code can be obtained, the troubleshooting charts will not be too effective if the check engine light is off. When the check engine light is off, it means that the system is operating normally and, consequently, any troubleshooting tests that are made will most likely

79

Introduction to Automotive Solid-State Electronics

**DRIVER COMPLAINT OR EMISSION FAILURE
ENGINE PERFORMANCE PROBLEM
(ODOR, SURGE, FUEL ECONOMY...)**

- Cold operation complaint.
  2.8L V-6 See Chart #6 - lean limit switch check.
- Full throttle performance complaint
  3.8L V-6 See Chart #4 for 3.8L V6
  2.5L L-4 See Chart #4 for 2.5L L4 (Cold performance complaint only).
  2.8L V-6 See chart #4 for 2.8L V6
- All other complaints - follow Chart below on warm engine (upper radiator hose hot).

1. Place transmission in park (A.T.) or Neutral (M.T.) and set park brake.
2. Start engine
3. Ground trouble code "test" lead.
4. Disconnect purge hose from canister and plug hose.
5. Connect tachometer.
6. Disconnect mixture control (M/C) solenoid and ground M/C solenoid dwell lead.
7. Run engine at 3,000 rpm and while keeping throttle constant, reconnect M/C sol. and Note RPM.

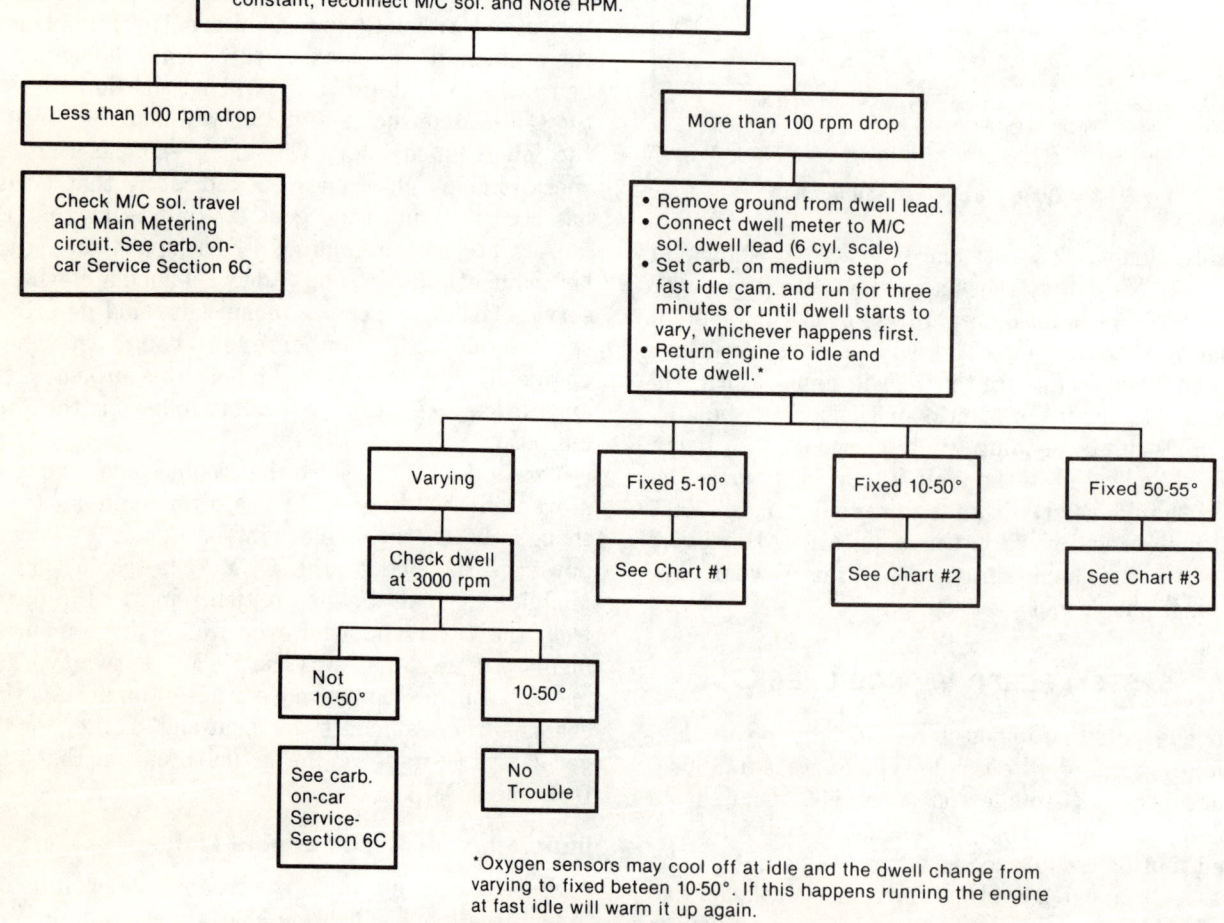

Less than 100 rpm drop
Check M/C sol. travel and Main Metering circuit. See carb. on-car Service Section 6C

More than 100 rpm drop
- Remove ground from dwell lead.
- Connect dwell meter to M/C sol. dwell lead (6 cyl. scale)
- Set carb. on medium step of fast idle cam. and run for three minutes or until dwell starts to vary, whichever happens first.
- Return engine to idle and Note dwell.*

Varying → Check dwell at 3000 rpm
- Not 10-50° → See carb. on-car Service-Section 6C
- 10-50° → No Trouble

Fixed 5-10° → See Chart #1

Fixed 10-50° → See Chart #2

Fixed 50-55° → See Chart #3

*Oxygen sensors may cool off at idle and the dwell change from varying to fixed beteen 10-50°. If this happens running the engine at fast idle will warm it up again.

The system performance check should be performed after any repairs to this system have been made.

Courtesy GM Product Service Training, General Motors Corporation

Fig. 7-5. System performance check procedure.

On-Board Diagnostic Systems

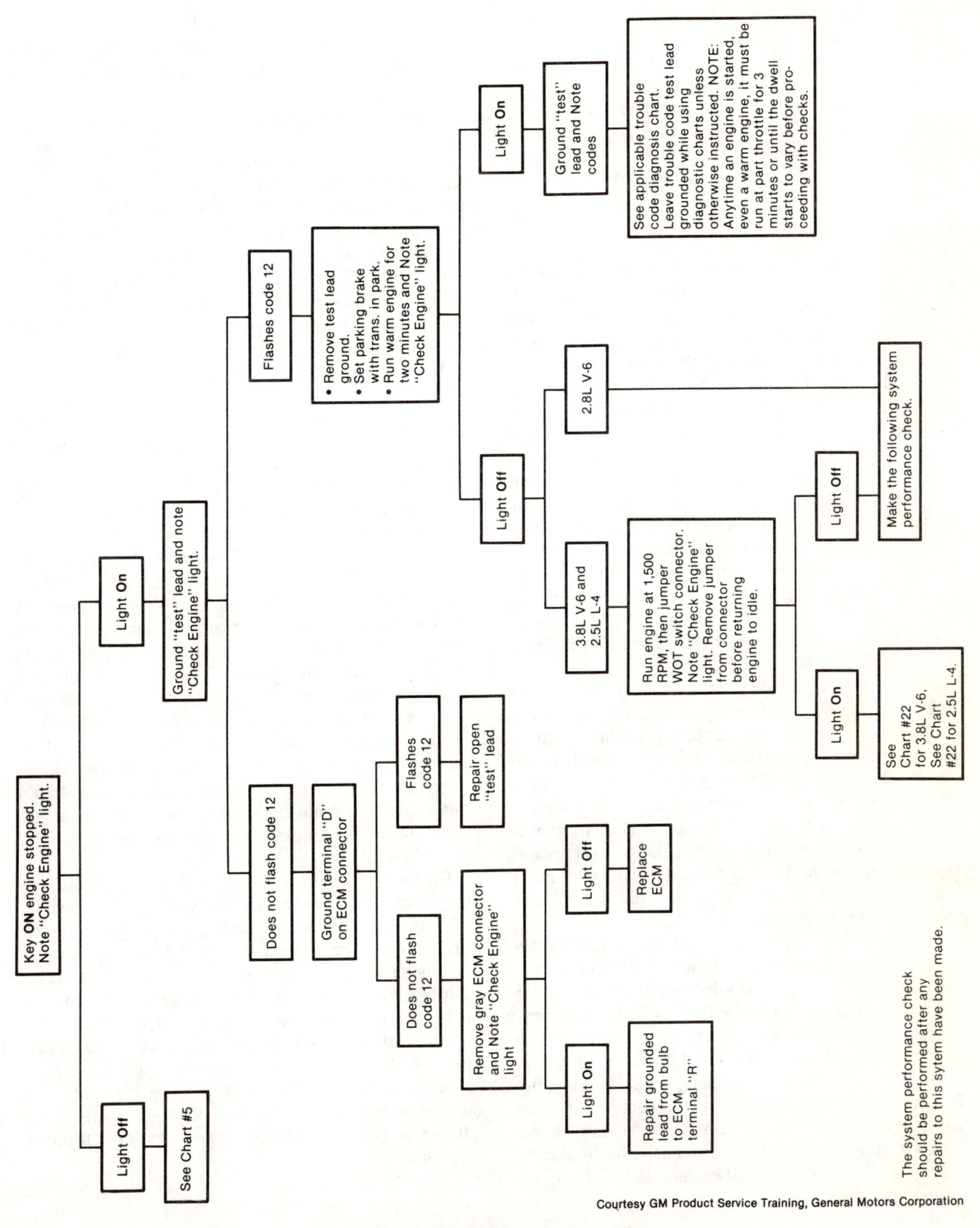

Fig. 7-6. Diagnostic circuit check procedure.

Courtesy GM Product Service Training, General Motors Corporation

Introduction to Automotive Solid-State Electronics

indicate normal operation. About the only thing that can be done in this situation is to make a thorough visual check of the components listed in the troubleshooting chart for the particular trouble code. Although this is not as certain a procedure as we would like, it is certainly better than no indication at all. And this is what would happen if the ignition were turned off before the trouble code was retrieved.

Memory Options

To get around this dilemma, the C-4 system offers two types of memory options. The first is the *short-term* memory which was just described. It holds the trouble code (even though the check engine light may have gone off) for as long as the engine is running. This is the operating mode as the vehicle is delivered.

The second option is the *long-term* memory which the service technician may initiate whenever the need arises. Under this memory option, whatever trouble code (or codes) is stored in memory remains in memory even after the ignition is turned off. This means that the driver need not bring his vehicle in for service immediately after the check engine light goes on. He can stop and start his engine as often as he pleases without losing the data stored in memory. To recover the trouble code from memory, it is necessary only to ground the test lead from the ECM as described previously.

This is a powerful diagnostic tool for the service technician and one which can go a long way in helping to solve those elusive intermittent problems. Bear in mind that, even if a trouble code is obtained, if the check engine light is off during the time the system is being tested, the individual components will probably test normal. Although this is somewhat of a hindrance to troubleshooting, at least the general area of the intermittent problem is defined; this is a lot more information than would have been available without the long-term memory.

How is the long-term memory activated? In the ECM there is a terminal marked S. By connecting this terminal directly to the battery (or to any other point that remains permanently "hot"), the memory-storage circuits will be kept on, even if the ignition switch is turned off. However, this places a continuous drain on the battery. Although the drain is quite small, and would have no effect on a vehicle in normal service, it could result in a run-down battery for vehicles in storage. Also, there is no need for long-term memory unless an intermittent problem is being diagnosed. For this reason, it is important that the memory be deactivated (S terminal disconnected) after the problem has been solved and corrected.

ADDITIONAL DIAGNOSTIC TECHNIQUES

A few additional remarks on the C-4 system. Tachometers can be connected to this system in much the same manner as other HEI systems (i.e., from the tach terminal to either ground or the hot post of the battery, depending on the tach manufacturer's instructions). However, the tach terminal at the distributor is not directly available on C-4 systems since this terminal is used to send engine-speed data to the ECM. By tracing the tach wire back from the distributor, a two-terminal connector will be encountered with one unused connector. Use this connector for the tachometer but do not disconnect the original connections since this will prevent the ECM from receiving tach data.

Inasmuch as the ECM is basically a computer, it must have a program to operate it. This is contained in a separate calibration unit called a PROM, which plugs into the ECM. PROM stands for Programmable Read Only Memory and contains all of the instructions needed by the computer to evaluate the input data and output the proper control signals for a particular engine application. A different engine would require a different PROM. Thus, by utilizing the same basic computer (the ECM), the engine manufacturer need only change the PROM to accommodate different engine requirements. Unlike the "volatile" memory used to hold the trouble codes (which remembers only as long as power is applied), a PROM has a *static* memory: the program it contains has been burned in and can never change, whether power is applied or not.

On-board diagnostics are not confined only to engine-control systems, they can be applied to virtually any system incorporating a computer (microprocessor). For example, the Electronic Message Center used on various Ford products has just such a self-testing feature. Like the C-4 system, this one too is programmed to flash a series of data, although in this case the data appears on an alphanumeric display panel—a readout device that shows both alphabetical as well as numerical characters. The basic mode of operation is shown in Fig. 7-7. Note that this procedure also relies on additional tests and troubleshooting charts.

On-Board Diagnostic Systems

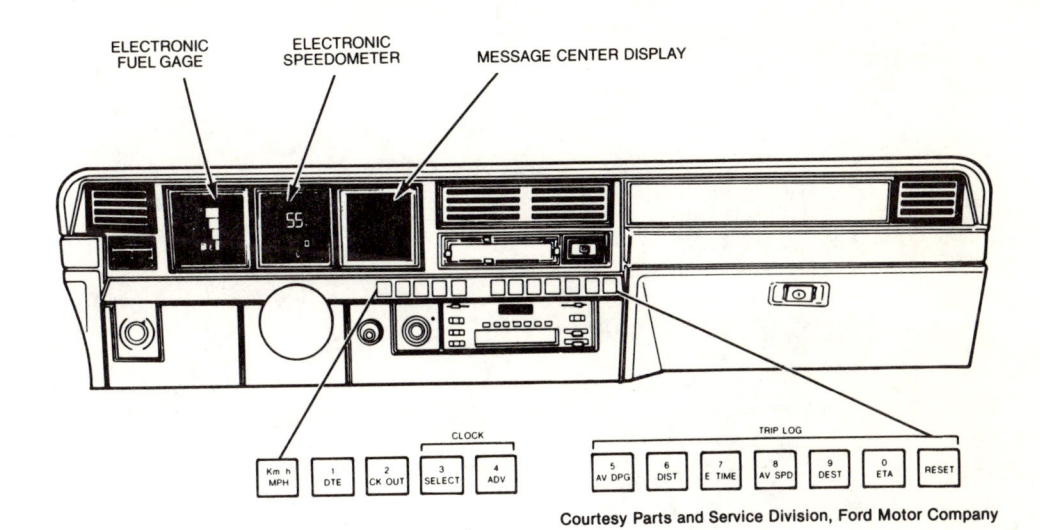

Fig. 7-7. Ford diagnostic system utilizing its Message Center System.

Courtesy Parts and Service Division, Ford Motor Company

CHAPTER 8

Electronic System Developments

The rapid growth of solid-state electronics, particularly in the area of microprocessors and microcomputers, has not only changed the approach to engine control but has spawned a whole new array of auxiliary vehicle systems. Some of them are probably more in the realm of frills and gadgets, but others offer improved driving safety and comfort.

Here is a brief review of some of the more noteworthy developments. We will not go into much detail on how they function because some of the systems are quite complex despite the simplicity of their purpose. In the future we can expect to see many changes, deletions, and additions to this group of accessory systems.

KEYLESS ENTRY SYSTEM

In an effort to minimize auto theft and/or make illegal entry into an automobile more difficult, Ford offers a Keyless Entry System in some of its luxury models. Outwardly, all one sees is a five-button keyboard on the outside panel of the driver's door. At first glance, it appears that ten keys are involved (see Fig. 8-1), but in reality each key represents two numbers. This keyboard feeds a microcomputer/relay module that is usually located under the hood.

The computer has been programmed at the factory with a five-digit code number which the owner, hopefully, has memorized. In case he forgets, it is recorded on his owner's warranty card and also on the inside lid of the trunk. Rather than use some arbitrary code of the factory's choosing, the owner can reprogram the computer to accept any other five-digit code number that has special meaning for him, such as a birthdate, phone number, etc. Although not touted in the manufacturer's literature, a certain degree of sobriety is needed to make the system work.

When the proper sequence of digits has been entered into the keyboard, the following will happen: (1) the driver's door unlocks; (2) the interior lights go on if the 1/2 key is pressed; (3) all other doors unlock if the 3/4 key is pressed within 5 seconds after the driver's door unlocks; (4) the rear decklid (trunk) unlocks if the 5/6 key is pressed within 5 seconds after the driver's door unlocks.

All doors will automatically lock when (1) the driver's seat is occupied (except those with Recaro seats), (2) all the doors are closed, (3) the ignition is turned ON, and (4) the transmission is in R, N, D, or L. All doors can be locked from the outside by pressing the last two buttons (7/8 and 9/10) simultaneously.

The modules, which are programmed at the factory, have a five-digit number imprinted on them which refers not to the keyboard numbers but to the button locations. If for any reason the access code number should be forgotten, the warranty card lost, or the trunk sticker damaged, the access code can be derived from the five-digit number on the module. Should it ever be necessary to replace the module, a new code number (printed on the new module) must be used. As you can see, the same module number can generate a variety of keyboard code numbers all

Fig. 8-1. Keyboard for the Keyless Entry System.

Courtesy Parts and Service Division, Ford Motor Company

of which will work. An overall view of the circuit is shown in Fig. 8–2.

KNOCK LIMITERS

It has long been known that improved engine efficiency occurs when the engine operates just below the threshold of "pinging" or "knocking." The ignition timing is occurring at the near-optimum point in the combustion cycle, meaning that the burning air–fuel mixture can develop its maximum thrust on the piston.

However, when ignition takes place too soon, combustion pressures build up too rapidly, causing the as yet unburned portion of the mixture to suddenly self-detonate. This leads to excessively high combustion pressures that actually fight the motion of the pistons. It is not unlike the inexperienced bicycle rider who applies pressure on the pedal before it passes top dead center: even though the force is there, it does not do much. In an internal combustion engine the result of improper ignition timing is an audible "knock." Not only does it lessen the efficiency and power of the engine, it places unduly high stresses on the piston and its associated parts.

The ideal situation would be to operate the engine as close to the point of knocking as possible, but without actually knocking. By combining the computational speed of a microprocessor with a knock or "detonation" sensor, this ideal can now be closely approached.

A characteristic of engine knock, even at fairly low levels, is the particular sound frequency it generates. The knock limiter, or detonation sensor as it is also called, is tuned to this frequency. When it detects this frequency, it sends a low-voltage signal to the spark control computer (microprocessor). The computer analyzes the strength and frequency of this signal, according to the program stored within it, and retards the timing a certain number of degrees. When the knock is no longer detected, timing returns to its normal setting, which is controlled by the other inputs to the computer.

Fig. 8–3 shows a typical installation with the knock limiter located on the intake manifold. It should be remembered that the knock limiter is but one of several sensors feeding data to the computer that controls ignition timing. Not all systems use such a sensor; but on those that do, timing can be fine-tuned, under actual driving conditions, for a high degree of efficiency.

The functioning of such a device can be demonstrated in the following manner. Connect a variable delay timing light (the type used for testing timing advance) to the engine. Run the engine at a fast idle (approximately 1200 rpm) and, using the timing light, bring the timing mark into view. Then tap lightly on the manifold near the sensor with a small metal object. When you do this, you should be able to note a decrease in the spark timing. The amount of decrease is directly proportional to the strength and frequency of the tapping. The maximum amount of timing retardation is controlled by the computer program (11° might be a typical value).

WIPER SYSTEMS

Windshield wiper systems are usually simple affairs consisting of a variable or two-speed drive motor. A somewhat more sophisticated system involves the use of an electronic time-delay circuit to provide intermittent wiper operation. This type of operation is preferrable under some misting conditions when regular and rapid wiper motion might smear the windshield.

Fig. 8–4 shows the circuit of a typical wiper delay system. The heart of the system is a so-called Schmitt-trigger circuit that can be adjusted, by the driver, to provide a range of time delays. At the end of each time-delay period, the wiper makes one sweep and the delay is reset. A time-delay circuit works by charging a capacitor through a resistor; the greater the value of the resistance, the longer it takes the capacitor to charge. When the capacitor reaches a certain level of charge, its voltage is high enough to trigger the circuit. At this point, the wiper motor is momentarily turned on and the capacitor is discharged. The circuit then begins its next cycle of operation.

ANTISKID BRAKING

Any experienced driver knows that excessive braking on slippery roads causes the brakes to "lock-up," resulting in virtually no braking reaction at all. The driver also knows that rapid pumping of the brake pedal is an effective way of slowing down a vehicle under slippery conditions. In other words, an attempt is made to apply a braking force that is just under the threshold of lock-up and skidding. How well the driver succeeds depends on individual driving skill and presence of mind; the less skillful tend to panic, apply too much brake pressure, and go into an uncontrolled skid.

Antiskid braking systems seek to emulate, with greater precision, just such a braking technique.

Electronic Systems Developments

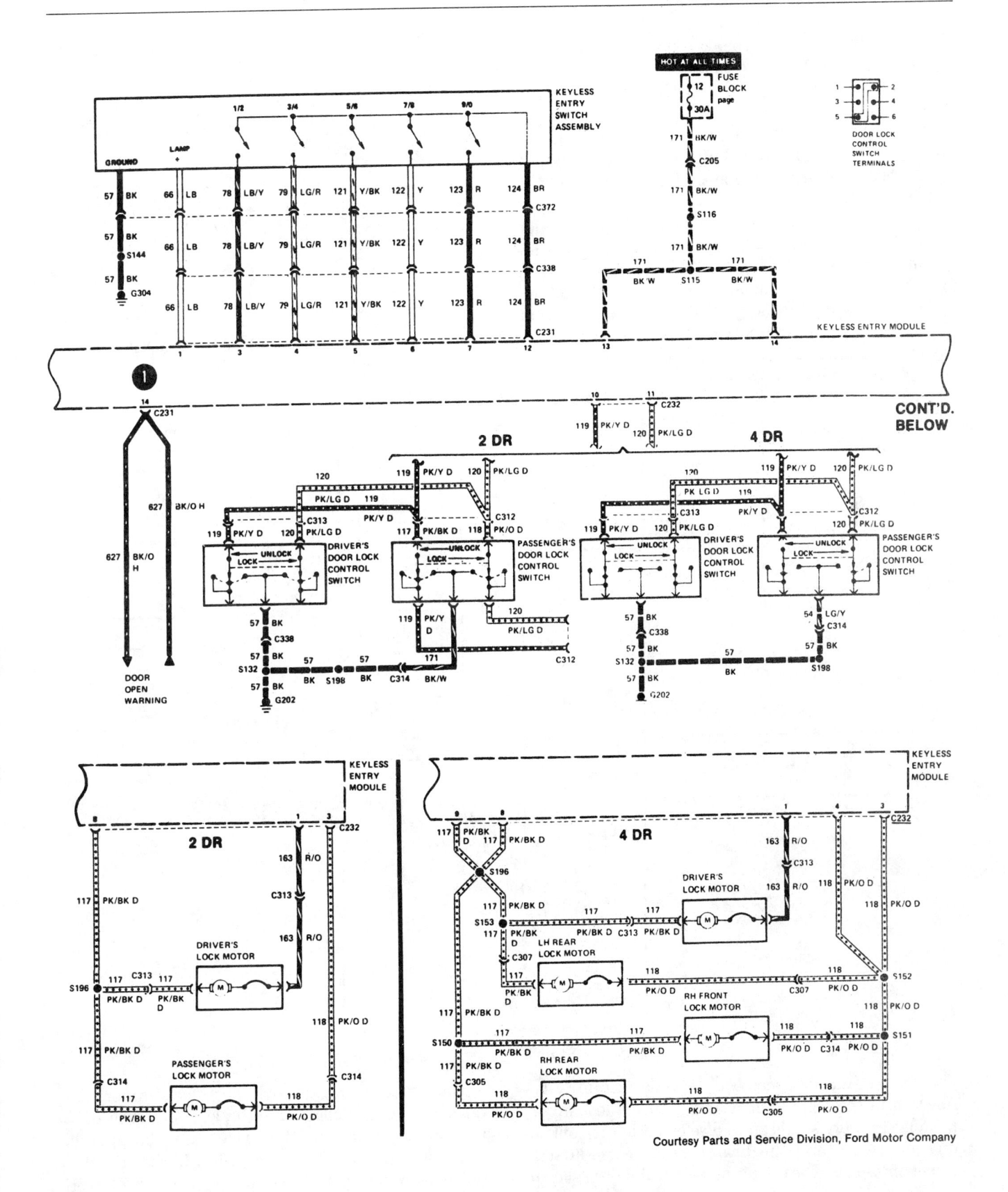

Fig. 8-2. Schematic of the Keyless Entry System.

Courtesy Parts and Service Division, Ford Motor Company

Introduction to Automotive Solid-State Electronics

They accomplish this through a system of electronic sensors and pneumatically controlled brake actuators, or modulators as they are sometimes called. These systems fall into two general classes: two-wheel systems and four-wheel systems. The four-wheel systems are more effective but also more costly. In two-wheel systems, antiskid braking applies only to the rear wheels.

The basic control system consists of three major elements: (1) wheel-speed sensors, (2) brake-pressure modulators, and (3) a logic control unit. The purpose of the wheel-speed sensors is to detect sudden *changes* in wheel velocity. This is the indicator of brake lock-up or skidding. The decision as to just how

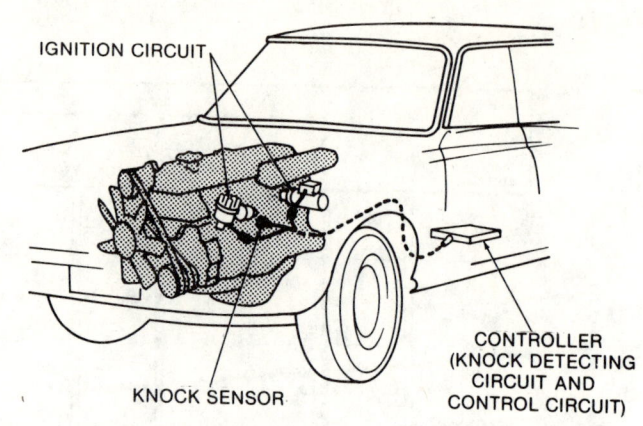

Fig. 8-3. Typical knock sensor installation.

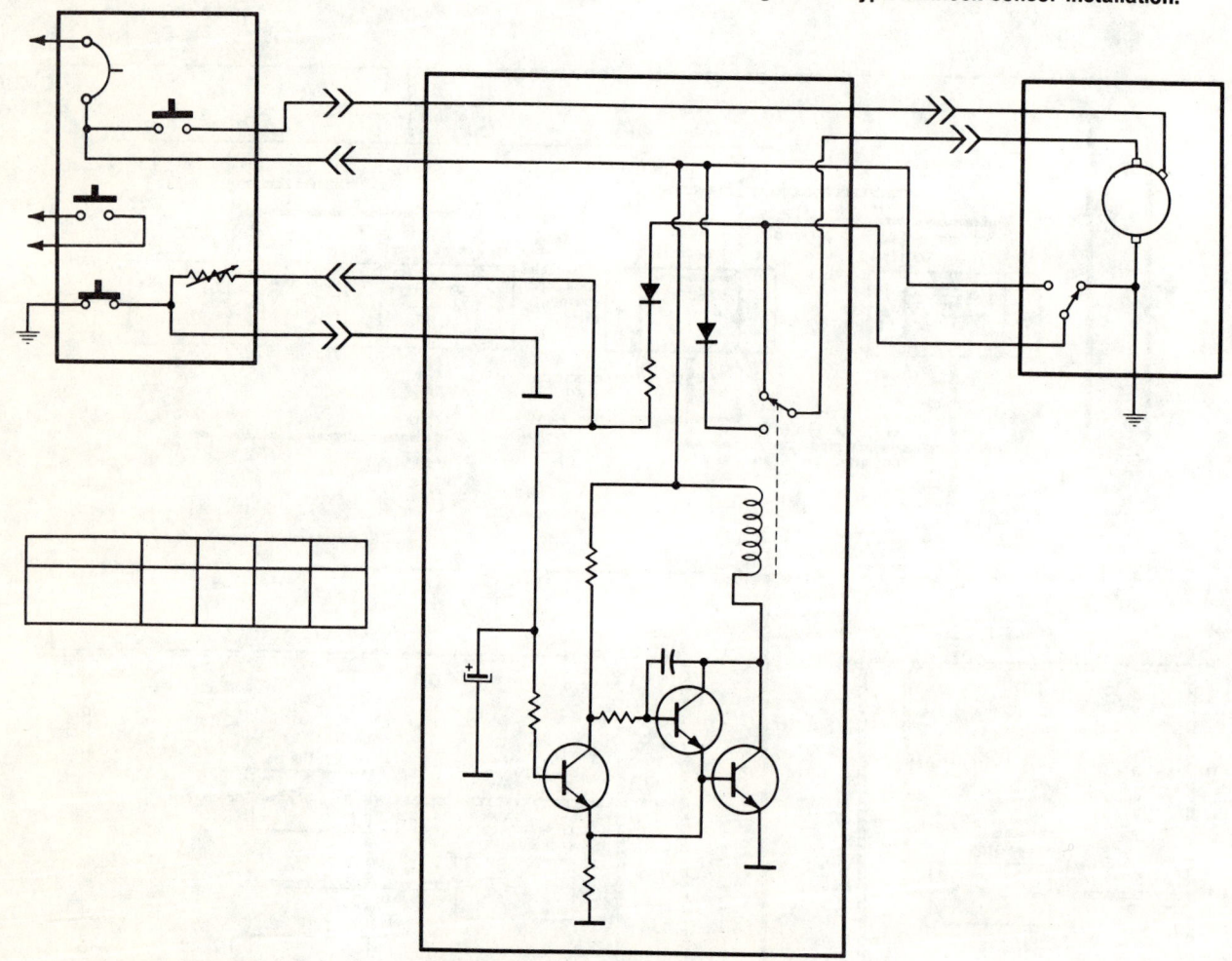

Fig. 8-4. Diagram of a wiper delay system.

much of a change in speed constitutes a lock-up is the function of the logic control unit. When a lock-up or pending lock-up is indicated, a solenoid in the control unit energizes and the brake-pressure modulator comes into play. The circuit of a control unit is shown in Fig. 8-5.

The brake modulator, which is vacuum operated, takes over the application of braking pressure to the wheels, even though the driver may be applying heavy pressure to the brake pedal. It does this by a series of rapid on-off pressure applications directly to the hydraulic fluid in the brake lines. (In a two-

88

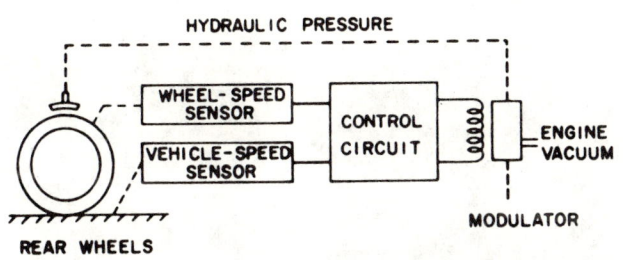

Electronic Systems Developments

Fig. 8-5. Basic block diagram of antiskid braking system.

wheel system, of course, it would do this only to the rear brakes.) See Fig. 8-6 for a cross section of a typical brake modulator. If the wheel-speed sensors signal that a lock-up is still occurring, the modulator releases hydraulic pressure until the wheel is again moving at road velocity. At this point, the modulator reapplies pressure until lock-up is again indicated. This process can repeat itself many times a second, always under control of the logic unit, so that the overall effect is that the brakes are being applied just below the point of skidding.

The net result is that the brakes are being applied at the maximum pressures compatible with the "slipperyness" of the road—regardless of the brake pressure being applied by the driver. However, when the driver releases the brakes, the antiskid system ceases to function. Also, if the driver applies braking pressure below the skid point, the system does not take over the braking action.

The wheel-speed sensors are electromagnetic devices not unlike those used as the triggering elements in electronic-ignition systems. They are not always mounted on the wheels. In a two-wheel system there may be only one sensor receiving its signal from the drive shaft. In this case, the sensor is sensing the average speed change of both wheels. Four-wheel systems usually employ sensors at each wheel and may use as many as four modulators.

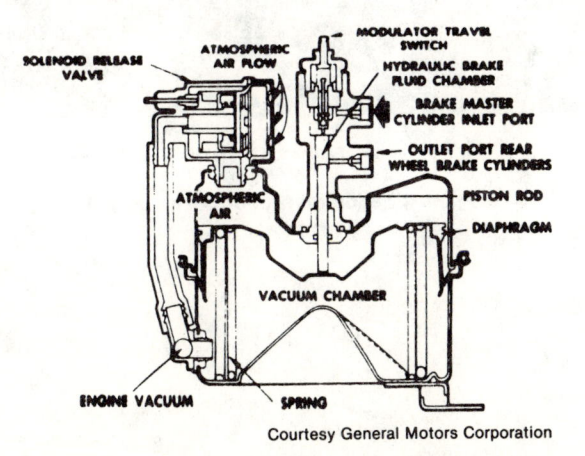

Fig. 8-6. Cross section of antiskid brake modulator.

The future of antiskid braking does not necessarily end here. Considerable work is being done on a much more sophisticated system employing radar detection of vehicle speed. When coupled with the wheel-speed sensors discussed above, it will offer even better detection of wheel lock-up or skidding. The ultimate braking system, as some propose, will be the application of radar control to automatic braking to prevent rear-end collisions and similar accidents. In such a system, radar will constantly monitor the rate at which a vehicle is approaching another object in its path. If the relative speed differential is too great or the separation distance too little, the brakes will automatically be applied.

TRIP COMPUTERS

The microprocessor has opened up a whole new area of automotive instrumentation that would have been virtually impossible (or economically impractical) a few years ago. These are the trip computers, the computer-controlled dashboard instruments that give drivers useful—and sometimes trivial—information on the progress of a journey. Some, like Ford's Message Center, provide important data as to the condition of vital engine and safety systems.

Such systems consist of three main assemblies: (1) a display module that serves as the data readout, (2) a keyboard through which the driver can input data and commands, and (3) the control or computer module itself. The latter is usually mounted behind the dash. A typical system as seen from the dashboard is shown in Fig. 8-7. Here are the basic functions such systems perform.

Miles per Gallon—(Not available on all vehicles.) This shows the average fuel consumption over a brief period of time (e.g., 4 seconds). The system monitors the quantity of fuel consumed over the time period as well as the average vehicle speed. Using the arithmetical ability of the computer, it divides the latter by the former and converts the result into miles per gallon. The result appears on the display.

Trip Speed—This is not quite the same as the speedometer reading. It is the average speed maintained by the vehicle since the last time the driver keyed in this function. Average speed is a better indicator of trip progress than the instantaneous reading of the speedometer.

Trip Time—This function displays the total elapsed time since the beginning of the trip. During the first hour it shows this in minutes and seconds; afterwards, it shows it in hours and minutes.

Range or Distance to Empty—By measuring the

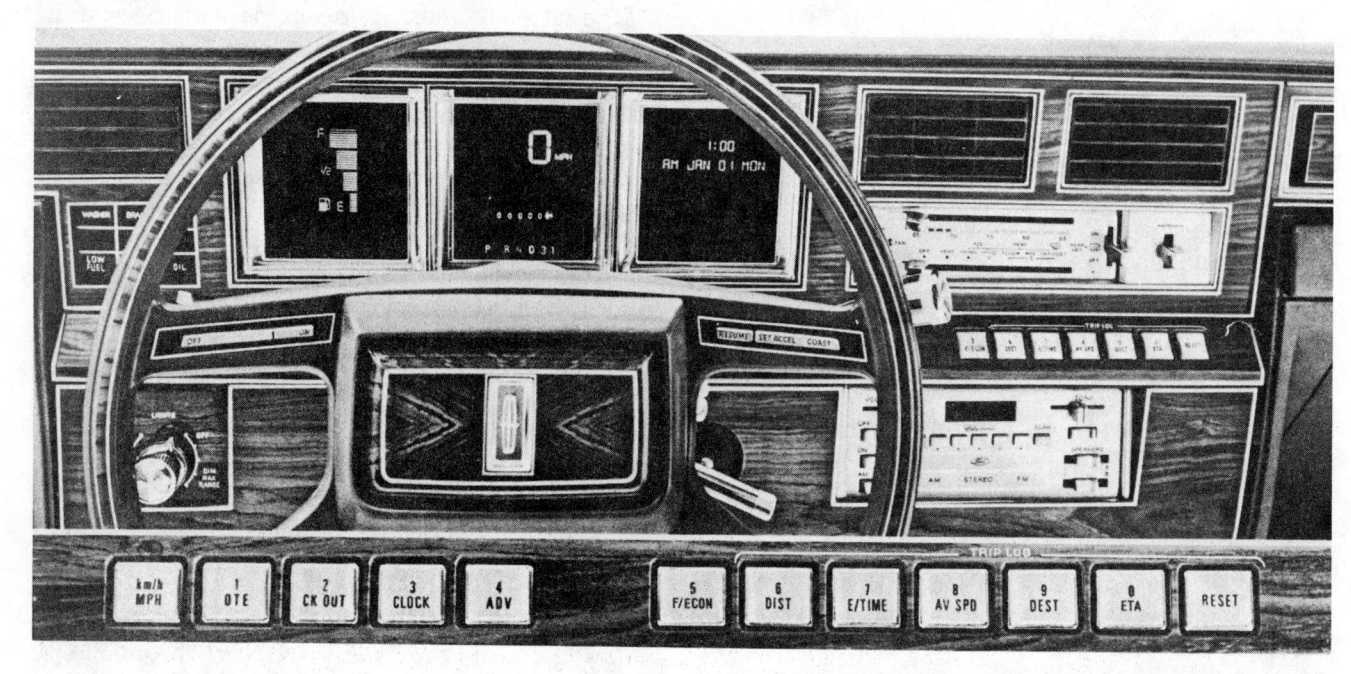

Introduction to Automotive Solid-State Electronics

Fig. 8–7. Trip computer on dashboard.
Courtesy Parts and Service Division, Ford Motor Company

amount of fuel remaining in the tank and computing the average fuel consumption, the system can calculate the number of miles of driving left in the tank. This calculation is updated every 4 seconds to accurately show changes in driving conditions.

Destination—If the driver has entered initially the total number of miles for the trip, the computer will subtract the miles already driven and display the result—the number of miles remaining.

Estimated Time of Arrival—Similar to the destination function, this one shows the estimated time of arrival based on the average trip speed, the total miles, and the current time. This assumes, of course, that the driving conditions remain fairly consistent.

In addition to trip data, these systems can also provide such information as engine rpm, coolant temperature, system voltage, low-fuel warning, low washer fluid, low brake pressure, headlamp out, taillamp out, brakelamp out, trunk ajar, door ajar, low oil pressure, and alternator condition. Any one system may not provide all of these indications, but a good share of them. As we discussed in Chapter 7,

some of these systems contain a self-diagnostic program to help localize the problem should they ever develop trouble.

Closely associated with trip computers are the electronic versions of the standard dashboard instruments. These are usually digital or graphic displays for fuel and speedometer readout. See Fig. 8–8. One of the advantages of a digitally displayed speedometer is that it can be programmed to readout in either miles per hour or kilometers per hour simply by flipping a switch.

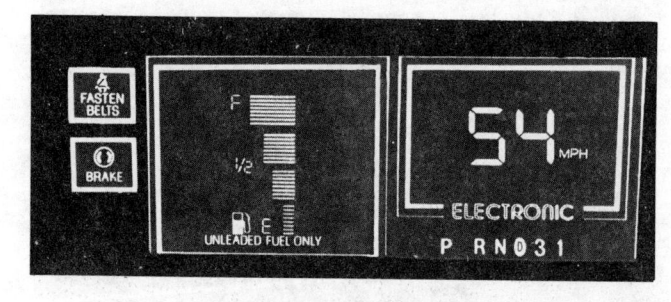

Courtesy Parts and Service Division, Ford Motor Company

Fig. 8–8. Digital speedometer readout and graphic fuel level display.

Index

A

Address, 69
AFR; *see* air fuel ratio(s)
Air-
  fuel
    mixture, 86
    ratio(s), 32, 45, 46, 69, 70
  injection pump, 47-50
Alphanumeric display, 82
Alternator(s), 19, 22-26
  basic, 22-23
  condition, 90
  diode use, 15-16
  improvement of basic, 23-26
  single-phase, 23
  three-phase, 23
Ambient temperature, diode, 16
Ampere, 7
Amplifier,
  transistor, 17, 18
  vacuum, 51
AND gate, 61
Anode, diode, 15
Antiskid braking, 86
Armature, 12
  relay, 27
Automatic lamp controller, 18-19
Automotive
  engine combustion, 45-46
  logic units, application of, 64-66
Avalanche mode, 29

B

Ballast resistor, 7, 35
Base
  current, control, 18
  transistor, 17-18
Battery
  charging, 21
  construction, 21-22
  failure, 22
  lead-acid, 21
Beta, 18
Bimetallic hinge, 28
Binary system, 67-68
Bit, 69
Brake-pressure modulator, 88-89
Braking, antiskid, 86
Branches, parallel circuit, 10

Breakdown point, diode, zener, 29
Breaker-point(s), 36
  ignition, 34
  systems, 35
Byproducts of incomplete combustion, 45
Byte, 69

C

Cap, filler, 48
Carbon monoxide, 45, 70
Carburetor icing, 52
Catalyst, 54
Catalytic converters, 54-55
  honeycomb type, 54
  monolithic, 54
  pellet type, 54
Cathode, diode, 15
Charcoal canister, 47
Charging
  and discharging processes, 21
  system, generator-driven, 26
    current-limiter, 26-27
    cutout relay, 26
Check engine light, 78, 79-82
Choke system, electric-assist, 52
Circuit(s)
  check, diagnostic, 79, 81
  complete, 8
  electrical, 7
  logic
    AND, 61
    EOR, 64
    NAND, 64
    NOR, 64
    NOT, 63
    OR, 62
  types of, 9-11
    parallel, 10
    series
      -parallel, 10-11
      voltage drop in, 9-10
CO; *see* carbon monoxide
Codes
  fault, 78
  trouble, 78
Coil
  field, 26
  output reduction, solving, 35
  saturation time, 34-35

Index

Collector, transistor, 17–18
Combinations of logic gates, 64
Combustion
  automotive engine, in, 45–46
  incomplete, 45
  process, 32
Computer
  -controlled EGR valve, 50
  on-board, 66, 69
  spark control, 86
  trip, 89–90
Condenser, 34
Connection, bad, 10
Construction of the transistor, 17–18
Conventional current flow, 8
Converters
  blockage, 55
  catalytic, 54–55
Current, 7
  flow
    conventional, 8, 15
    electron, 8
Cutoff, 19

D

Data storage, 68–69
Delta connection, 24
Destination, 90
Detonation sensor, 86
Diagnostic
  circuit check, 79, 81
  subsystem, 78
Diamagnetic, 11
Dieseling, 53
Digital display, 90
Diode
  assembly, 25
  connections
    forward bias, 25
    reverse bias, 25
  elements, 15
  heat
    dissipation, 16
    sinks, 16–17
  ratings, 16–17
  reverse state, 17
  uses, 15–16
  zener, 29–30
Displays
  fuel, 90
  speedometer, 90
Draft tube, 46
Dual-diaphragm EGR valve, 51
Dwell, 35
  meter, 78, 79

E

ECM; *see* electronic control module
EGR; *see* exhaust gas recirculation
Electric-assist choke system, 52

Electrical
  circuits, 7
    solid-state, 15
    types, 9–11
  potential, 8
  pressure, 7
  terms, 7
Electrolyte, 21, 22
Electromagnetic induction, 22
Electromagnetism, 12–14
Electromechanical voltage regulator, 27–28
Electromotive force, 7
Electron current flow, 8, 12
Electronic
  control module, 79, 82
  ignition systems
    troubleshooting data, 38–43
      American Motors, 38
      Chrysler, 39
      Delco (GM) (HEI), 40
      Ford, 41–43
  Message Center, 82
  spark timing, 71–72, 74
Electrons, 7
Elements
  diode, 15
  transistor, 17
Emitter, transistor, 17–18
Engine, automotive, combustion, 45–46
EOR gate, 64
EPROM; *see* erasable PROM
Erasable PROM, 69
EST; *see* electronic spark timing
Estimated time of arrival, 90
Evaporation control systems, 47
Excessive resistance, 10
Exclusive OR gate, 54
Exhaust gas recirculation, 46, 50

F

Feedback carburetors, 69–71
Ferromagnetic materials, 11
Field
  coils, alternator, 12, 26
  current
    control, 26
    switching, 30
  relays, 28
Flux density, 12
Force
  electromotive, 7
  lines of, 12
Formula
  Ohm's law, 8
  parallel resistors, 10–11
  power, 16
Forward current rating, 16

G

Gain, 18
Gates, logic
  AND, 61

Index

Gates, logic—cont
   EOR, 64
   NAND, 64
   NOR, 64
   NOT, 63
   OR, 62
Germanium, 17, 18
Ground
   circuit, 10
   side, circuit, 8

H

Hardware, 66, 67
HC; *see* hydrocarbons
Headlight circuit, 10
Heat
   dissipation, 16
   sinks, 16–17
     transistor, 19
High
   Energy Ignition system, 35
   -voltage energy production and limitations, 34–35
Hinge, bimetallic, 28
Honeycomb type converter, catalytic, 54
Hydrocarbons, 45, 46

I

I, symbol, 7
Idle stop solenoid, 53
Ignition
   circuitry, 31
     primary, 31
     secondary, 31
   coil, 12, 23
   demand, 35–36
   misfiring, 32
   problems, 32–33
   reserve, 36
   system, primary circuit, 7
Incomplete combustion, 45
Induction, electromagnetic, 22
Instructions, 67
Integral charging system, 28
Intermittent problem, 79, 82
Inverter, 62
Ion, free, 32
Ionization, 31–32, 33

K

Keyless entry system, 85
Knock
   detector, 37
   limiters, 86

L

Law, Ohm's, 7
Left-hand rule, 12
Light, check engine, 78
Limiter, knock, 86

Lines of force, 33
Load, 8
Loadstone, 11
Lock-up, brake, 86, 89
Logic
   control unit, 88
   decision making, and, 61–66
   function, 61
   gate(s), 61
     combinations of, 64
   systems and the computer, 66
   units, 64–66
Long-term memory, 82
Look-up tables, 69

M

Magnetic
   field, 12, 22–23, 33, 34
   flux, 22
   poles, 11–12
   properties, 11
   switches, 12
Magnetism
   artificial, 11
   natural, 11
Matter, magnetic properties of, 11
Memory
   location, 69
   options, 82
   volatile, 82
Message center, electronic, 89
Microcomputer/relay module, 85
Microprocessor, 66–67, 69, 75, 86
   and the automobile, 66–69
Microsecond, 19, 68
Miles per gallon, 89–90
Mixture-control solenoid, 70–71

N

NAND gate, 64
Nitrogen, formation of oxides of, 46
NOR gate, 64
NOT gate, 63
Npn transistor, 17–18
Numbers, binary, 67–68

O

Ohmmeter, 8, 18, 19, 78
Ohm's law, 7, 8–9
On-board computers, 69, 72
OR gate, 62
Orifice spark advance control, 53
OSAC; see orifice spark advance control
Oxides, nitrogen, 46
Oxygen sensor, 45, 54, 69–71
Ozone, 46

P

Palladium, 54
PAN, 46

Index

Paramagnetic, 11
PCV; *see also* positive crankcase ventilation valve, 47
Peak inverse voltage, 17
Pellet type converter, catalytic, 54
Permanent magnets, 11–12
Photocell, 18
Platinum, 54
   electrodes, 70
Pnp transistor, 17–18
Positive
   crankcase ventilation, 46–47
   feedback, 19
Potential, 7, 8
Pressure-vacuum relief filler cap, 48
Program, the, 67–68
Programmable read only memory, 69, 82
PROM; *see* programmable read only memory
Pulse generator, 36
Pump, air-injection, 47–50
Purge valve, 47

R

R, symbol, 7
Radar control, 89
RAM; *see* random access memory
Random access memory, 69
Range or distance to empty, 90
Ratings, diode; 16
Read only memory, 66, 69
Rectifier assembly, 25
Regulator(s), 21
   setting, high, 22
   solid-state, 28–30
Relay(s), 12
   armature, 27–28
Reluctor, 37, 72
Resistance, 7
Resistor, temperature-sensitive, 30
Reverse
   leakage current, 17
   state, diode, 17
Rhodium, 54
ROM; *see* read only memory
Rotor, 8–9, 23–24, 25–26

S

Saturation, 19
Self-testing
   capabilities, 78
   features, 75–78
Semiconductor material, 15
Sensor
   detonation, 86
   electronic, 88
   input, 72
   oxygen, 45, 54, 70–71
   thermal, 52
   wheel-speed, 88–89
Series circuit, 9–10
Short-term memory, 79, 82

Silicon, 17, 18
Single-phase alternator, 23
Smog
   London-type, 45
   Los Angeles-type, 45, 46
Software, 67
   program, 66
Solenoid, 12–13
   idle stop, 53
   mixture-control, 70–71
Solid-state
   circuits, 15
   regulator, 28–30
   switches, 34, 36–37
Source, voltage, 8
Spark
   control, 69
      computer, 86
   -delay valve, 53–54
   gap, 32
   plug operation, 31–32
   timing, electronic, 71–72
Speed, trip, 90
Spring tension, 27
Starter motor, 12
   circuit, 7
Stator windings, 23, 24
Stepping motor, 71
Stoichiometric
   mixture, 45
   point, 45
   ratio, 70
Storage battery, 21
Subsystem, diagnostic, 78
Sulfate, lead, 21, 22
Sulfuric acid, 21, 22
Switches, solid-state, 34
Switching device, 31, 33, 36–37
Symbol(s), 7
   transistor, 17
System
   choke, electric-assist, 52
   control, evaporation, 47
   diagnostic, on-board
      closed-loop, 76
      open-loop, 76–78
   hardware, 66
   performance check, 79–82

T

Tank filler cap, special, 46
Temperature
   compensated, regulator, 28
   -sensitive, magnetic shunt, 28
   resistor, 30
   shunt, magnetic, 28
Terms, electrical, 7
Testing equipment, 78–79
Thermal cutoff, 51
Thermistor, 30
Thermostatically controlled air cleaners, 51–52

Index

Three
  -phase alternator, 23
  -way converter, catalytic, 54
Transformer, 33
Transistor(s), 17–19
  check using ohmmeter, 19
  elements, 17
  function, 18
  npn, pnp, 17–18
  ratings, 19
  states, 19
  switches, 61
  uses, practical, 18–19
Trip
  computer, 68, 89–90
  speed, 90
  time, 90
Trouble codes, 78
Troubleshooting, 72
  systems, ignition, electronic, 38–43
Truth table, 65

U

Unburned hydrocarbons, 45, 46, 48, 70

V

V, symbol, 7
Vacuum amplifier, 51
Valve
  control, OSAC, 53
  EGR
    computer-controlled, 50
    dual-diaphram, 51
    thermal cutoff, 51
    venturi-vacuum, 51

Valve—cont
  PCV, 46
  pressure relief, 50
  purge, 47
  roll-over, 47
  throttle, 75
  vacuum-control, 54
Venturi-vacuum EGR, 51
Vibrating voltage regulator, 27
Volatile memory, 82
Voltage, 7
  divider, 18, 30
  drop, 9–10, 16
  regulator, 26–30
    construction, 30
    electromechanical, 27–28
    operation, 30
    purpose, 27
    solid-state, 27, 28–30
    temperature-compensated, 28, 30
    thermistor, 30
    vibrating, 27
  source, 8
  zener, 29, 30

W

Wheel-speed sensors, 88
Windings, 24–25
Wiper delay, 86
Wye connection, 24

Z

Zener
  breakdown voltage, 29, 30
  diode, 29, 30